Transboundary Environmental Problems and
Cultural Theory

Also by Marco Verweij

CULTURE IN WORLD POLITICS (*co-editor with Dominique Jacquin-Berdal and Andrew Oros*)

Transboundary Environmental Problems and Cultural Theory

The Protection of the Rhine and the Great Lakes

Marco Verweij
Research Fellow
Max Planck Project Group on the Law of Common Goods
Bonn
Germany

Foreword by Mary Douglas

First published 2000 by
PALGRAVE
Houndmills, Basingstoke, Hampshire RG21 6XS and
175 Fifth Avenue, New York, N. Y. 10010
Companies and representatives throughout the world

PALGRAVE is the new global academic imprint of
St. Martin's Press LLC Scholarly and Reference Division and
Palgrave Publishers Ltd (formerly Macmillan Press Ltd).

ISBN 0–333–91563–1

This book is printed on paper suitable for recycling and made from fully managed and sustained forest sources.

A catalogue record for this book is available from the British Library.

Library of Congress Cataloging-in-Publication Data
Verweij, Marco.
 Transboundary environmental problems and cultural theory : the protection of the Rhine and the Great Lakes / Marco Verweij ; foreword by Mary Douglas.
  p. cm.
 Includes bibliographical references and index.
 ISBN 0–333–91563–1
 1. Water quality management—Rhine River. 2. Water quality management—Great Lakes. 3. Environmental protection—International cooperation—Case studies. 4. Transboundary pollution—Case studies. 5. International relations and culture. I. Title.
 TD295.R48 V47 2000
 363.739'4526'09434—dc21
                                                    00–034787

10 9   8   7   6   5   4   3   2   1
09 08 07 06 05 04 03 02 01 00

Printed and bound in Great Britain by
Antony Rowe Ltd, Chippenham, Wiltshire

For Gusta Couwenberg and Arend Verweij

Contents

List of Figures and Tables

Figures

Tables

Acknowledgements

This book is much better than it would have been if I had not received the intellectual opposition and emotional support of many people. The theoretical part of this study greatly benefited from critiques that were generously offered by Valérie de Campos Mello, Mary Douglas, Ernst B. Haas, Rob Hoppe, Eivind Hovden, Edward Keen, Marieke Kleiboer, Richard Rosecrance, Per Selle, Michael Thompson and Bertjan Verbeek. Earlier versions of the empirical chapters were helpfully scrutinized by Thomas Bernauer, Thomas Risse and Karen Smith. Highly useful reading tips and materials were given by Michael Barnett, Marc Galanter and Graham Wilson. Furthermore, André Nollkaemper suggested the choice of case studies (the environmental protection of the Rhine and the Great Lakes), while Jim Walsh made clear to me, over a cappuccino, how I could structure the book. Probably without realizing it, Michael Banks (for better or worse) once gave me the courage to write a theoretical book like this. I have also received illuminating feedback on the whole book from Frank Hendriks, Adrienne Héritier, Christer Jönsson, Gerd Junne and Jan Zielonka. Last, Dan Oakey improved the language of the entire manuscript. I am grateful to all these persons for their help.

I spent the summer of 1997 working in the Institute for International Studies at the University of Wisconsin, Madison. There I undertook my case study of the environmental protection of the Great Lakes. Dean David Trubek and his colleagues went beyond the call of duty in providing me with all the facilities that I needed to do my work. I am thankful for their hospitality.

During the course of my research, I interviewed 109 people representing organizations involved in the pollution and protection of the Rhine and the Great Lakes. I would like to express my gratitude to these people for having been so kind as to share their time and experiences with me.

Last, a word for my sponsors. As this research has been expensive, I could not have undertaken it without the financial help that I received from the Dutch Council for Scientific Research (NWO), the KLM Educational Foundation, the Ir. Cornelis Lely Stichting,

Dr C. L. van Steeden Fonds, Bekker-La Bastide Fonds and Vereniging Algemeen Studiefonds.

I dedicate this book to my parents, Gusta Couwenberg and Arend Verweij – with love.

Marco Verweij

Foreword

To compare the recent history of the Rhine and the American Great Lakes gives a rare vantage point for political theory. This book is a double case history of environmental protection in which both the river and the lakes have been very polluted, and in which the great river has finally done well but the great lakes not so well – and it is important to understand why. The writing has the style and control that draw the reader, who may never have given much thought to the Rhine, into its definitions and statements of the problem. I am proud to be invited to write a foreword, all the more because it is within the range of my interests, though outside my competence. I am not a political scientist but I know how to recognize judicious balance and systematic application of a theory, especially when I know the theory quite well.

One point I would like to clarify. In the book, Marco Verweij sometimes speaks of the 'cultural analysis of Mary Douglas' and of the 'Douglas school', suggesting that I had engendered it, or owned it, or invented it. I am pleased to be associated with the fast-growing and exciting field. Yet, I feel that this wording overstates my contribution, leaving those of others unjustly in the shadows. I will admit freely that I first had the idea that the social sciences should not remain steeped in the bias of their contemporary cultures. I do firmly believe that without a way of standing outside current conceits and commitments there could never be a place for anthropology in social thought. But what I had in mind was not a theory, but a method. That is why I called it an 'analysis' (grid-group analysis). It was to be a system for recognizing and explaining unconscious bias, a way of incorporating culture objectively into any account of social behaviour. It started out as an unfocused general approach that might be applied to all sorts of contexts. It is also true that I have written on political issues, but always in collaboration with an expert.

It was Aaron Wildavsky who saw how a method of studying cultural bias could generate theories about political prospects. And so he launched it and came to use the term 'cultural theory' for the specialized way of exploring politics that he and Michael Thompson had developed. In their hands it became a theory, with predictions, and with strong applications for public administration and political science. Early on they started to produce a distinguished series of publications, with the enthusiastic support of students at the Graduate School of Public Policy at Berkeley. Sadly, Aaron Wildavsky died before he could read this impressive book. Like a gift from a far away stranger, he would have welcomed it with surprise and joy, and with sorrow that he could not meet the author.

Wildavsky used to deny that he was a political scientist, or a political philosopher either. He like to describe himself as a 'policy analyst', indicating a modest journeyman's profession where common sense could be honoured and pretension rebuked. Before he came to cultural theory his fame and expertise were in budgetary studies. Over the last ten years of his life he was absorbed in the new interest and dedicated himself to a new objective, to develop a theory of political cultures. He would have rejoiced now to see how well his work has flourished. Verweij is not a lone swallow betokening the near arrival of summer – the cold weather has already retreated. In this book, he is in dialogue with other skilled practitioners of the art, and references abound in it to significant works that have appeared in the last five or so years, or are in press now. They are important studies, using the method of cultural theory and testing its assumptions on important political issues with weighty empirical materials.

Some of the monographs are by people whom Wildavsky had trained, and many by scholars he had never even met, like Verweij. Their contributions have the combined effect of removing a reproach that is often made against cultural theory (unjustly, as we vigorously protest), a complaint that the theory is not empirically grounded. To some degree the complaint is self-fulfilling, since the very academics who are standing there waving dismissively are the ones who could be setting their students to work to apply or develop the theory, or formally demolish it. This book is exemplary in this respect, showing how to make a tough, resilient argument, and suggesting what else might be examined.

For instance, it accords full weight to the institutional factors. Too often culture is taken to be an autonomous aggregate of personal attitudes, unmediated by institutions. But this mistaken idea can no longer be maintained. One of Verweij's innovations is to show how longer time depth enables hypotheses to be tested. Time reveals the effects that institutional structures have on personal attitudes. For example, most are ready to agree that political apathy (that is, fatalism, pessimism about the possibility of reform, a weak sense of subjective competence) is an enemy of democracy. Verweij's results suggest that apathy descends on a population that has become used over the long term to close regulation. He is not saying this in support of deregulationary policies, but he shows that the Rhine has been cleaned up thanks to the effective voluntary collaboration of major industrial agents, and he shows how solidarity in this direction was stimulated and kept alive. There is plenty of scope for more research to test, challenge or revise his conclusions.

What a long way has been travelled since the first programmatic statement in *Risk and Culture* (Douglas and Wildavsky, 1982[1]). River pollution figured large in political debates of the American 1960s and 1970s, which is where cultural theory started. The history of the pollution and clean-up of the Rhine remains at the heart of European politics, and now we have come full circle having gained much authority and confidence in this method of analysing pollution debates.

Mary Douglas

1
Introduction

A very brief history of the study of international relations

Some five to ten years ago, the study of international relations (IR) was flooded with calls to construct new theories and develop new paradigms. These calls were reactions to the perceived inadequacies of the IR theories that had been dominant since the early 1970s: neorealism, pluralism and structuralism.[1] In particular, numerous scholars argued that existing IR approaches had incorporated a number of false dichotomies that had reduced their explanatory and emancipatory power. The dichotomies that were singled out most often included the separation between the domestic/international realms, state/non-state actors, agents/structures, facts/values, idealism/realism and political/economic processes. It was extensively argued that most existing IR approaches had not provided an adequate analysis of their interactions. Many suggestions were made about how to provide new and better perspectives. We were urged to use our 'international imagination', to adopt sociological, anthropological, feminist and linguistic approaches, to turn to scientific realism and other 'post-positivist' epistemologies, as well as to ground our thinking in normative theory.[2] All this turned the study of international relations into a lively and exciting debate. However, not all was well. A major concern was raised by Thomas Biersteker in 1989. In his words:

> Up to this point, post-positivist scholars have been extremely effective critics but have been generally reluctant to engage in the construction and elaboration of alternative interpretations

1

or understandings ... I am not interested in a decisive proof of the superiority of post-positivism, but rather in a decisive demonstration of the plausibility of an alternative construction of some more concrete issue or subject.[3]

Biersteker's words very much resembled concerns raised by Robert Keohane, who argued that as long as the critics of traditional IR approaches lacked a clear empirical research programme they would remain at the margins of the discipline.[4]

During the last five years, these comments by Biersteker and Keohane have lost much of their validity. Many empirical analyses based on alternative approaches have come to light. Empirical research that has been informed by new frameworks has been undertaken on diplomacy, environmental cooperation, security regimes and foreign intervention (among other topics). Yet, in my view, something essential has been lacking in these empirical investigations: the development and testing of theoretical propositions. The empirical work that has been published during the last few years has tended to focus on the impact of highly specific ideas and perceptions in particular areas and eras. Very few attempts at generalization have been made. As a consequence, up to now, empirical analyses based on approaches other than those of the inter-paradigm debate have been neither cumulative nor always revealing.[5]

Why have only very few attempts at the development of theory been undertaken by the critics of neorealism, pluralism and structuralism? These critics can be roughly divided into two camps: poststructuralists and constructivists. Each group appears to have its own reasons. Poststructuralists have rejected the development of international theory on epistemological and normative grounds.[6] In their view, identities are what people make of them. National, gender and personal identities are in flux, are constantly being reinterpreted and renegotiated. As a consequence, it becomes impossible to formulate general theoretical propositions about international actors that are valid across time and space. Empirical analysis can therefore only reveal how identities are constructed in particular times and places.[7] This is an epistemological argument against the construction of theory. Poststructuralism also includes a normative objection to theory-building. A central part of the poststructuralist project is to reveal the internal tensions and contradictions that are hidden in any way of 're-presenting' the world, in any way of looking at the world, in any identity. According to poststructuralists, this serves an important emancipatory function. To them, the formation of any identity (e.g. conceptions of what it is to be a man, or to be Dutch) necessarily entails exclusion. Identities draw the line between those who are 'in' and those who are 'out'. As such, identities often form the basis for exclusionary, discriminatory practices. By revealing the contradictions and absurdities in the ideas and conceptions that make up identities, poststructuralists hope to delegitimize identities, and thereby to prevent discriminatory practices. This line of thinking gives poststructuralists a normative argument against theory-building. Academics engage in, and contribute to, the formation of identities when they attempt to formulate general propositions. For instance, in the past a number of IR scholars tried to explain international events in terms of 'race'. In doing so, they strengthened the belief that race is a useful concept with which to categorize people. But the concept has also often served as a basis for discrimination and subjugation of people.[8] Poststructuralists refuse to develop theoretical hypotheses as they do not want to reify contingent identities, thereby facilitating exclusion.

Fair enough. Poststructuralists do not want to contribute to the development of international theory, and can therefore hardly be faulted for not doing so. This does not apply, though, to the constructivists in the study of international relations. Constructivists have adopted many of the concerns of poststructuralists. They have also been interested in the (trans)formation of identities and its impact on international relations. They have also argued that international structures and processes are sustained by the ways in which actors regard themselves and others. But they have differed from poststructuralists in insisting that these ontological assumptions are compatible with theory-building. To them, the assumption that international relations are socially constructed does not necessarily preclude the development of theory. This stance has in fact made it possible for constructivists to seize IR's 'middle ground'[9] – the middle ground that had become void after neorealism, pluralism and structuralism had lost their central position in the study of international relations. Given the explicit aim of constructivists to develop international theory, it is somewhat surprising to note that their empirical analyses have had little theoretical content.[10] Constructivist empirical research has been by no means uninteresting, but has at the same time been quite specific and narrow in scope. As Jeffrey Checkel recently put it:

[C]onstructivists have convincingly shown the empirical value of their approach, providing new and meaningful interpretations on a range of issues of central concern to students of world politics. At the same time, constructivist theorizing is in a state of disarray. These researchers, much like the rational choice scholars they criticize, have made too rapid a leap from ontology and methods to empirics, to the neglect of theory development. This matters tremendously.[11]

I can only speculate about why constructivists have been so timid when it has come to developing theory. Perhaps it has something to do with the perceived failure of 'grand' theories, such as neorealism. Perhaps it is also related to unfamiliarity with cultural theories that have been formulated outside the study of international relations. Whatever the reasons, the logical next step in the constructivist project seems to be the development and application of theory. During the last few years, there have been a few signs of movement in this direction. Michael Barnett has applied the work of Erving Goffman in his research on Arab international politics. Ernst Haas has distinguished six basic forms of nationalism, each with a different impact on international relations, partly basing himself on the ideas of Ernest Gellner. Alexander Wendt has set out three 'cultures of anarchy' – Hobbesian, Lockean and Kantian.[12] However, a lot of work could still be done.

The aims of this research

The principal aim of this study is to contribute to this next step in constructivist theorizing about international relations. I agree with the constructivist criticisms of the three approaches from the interparadigm debate. And I neither endorse the epistemological assumptions nor fully share the normative stance of poststructuralism. But I also side with those who have criticized constructivism for its lack of theoretical content. With this book I attempt to introduce the discipline of international relations to a theoretical framework that has been developed in the fields of anthropology, sociology, political science and public administration over a number of years. My main claim will be that this framework meets a number of the ontological and theoretical standards that have been laid down by constructivists, while also avoiding some of the problems that have plagued constructivism so far. I will try to show this by applying the theory in empirical research on international environmental cooperation.

The framework that I will use is most often called cultural theory or (with less loftiness) grid-group analysis. The theory was first sketched out by social anthropologist Mary Douglas, and has been further elaborated into a general account of political and social life by Michael Thompson, Aaron Wildavsky, Richard Ellis and others. In brief, grid-group theory outlines four basic ways in which groups of people act and think (named *ways of life* or *cultures*). The theory assumes that these four ways of life are always represented (to a greater or lesser degree) in any area of social and political life. In fact, the cultural theory of the Douglas school sees social and political life as an ongoing competition between groups of people who adhere to one of these different cultures. Each of these social groups will attempt to impose, and argue for, its own way of living and perceiving. Within the fields of political science, anthropology and sociology, grid-group theory has already attracted wide attention. Its supporters have come to include some of the most distinguished present-day social scientists (including Harry Eckstein and Gabriel Almond), but so have its critics (e.g. Jeffrey Alexander, David Laitin, and Elinor and Vincent Ostrom). Until now, IR scholars have largely ignored grid-group theory. In this study, I will first extend the framework to the study of international relations, and then apply it in empirical research on transboundary environmental issues.

In my efforts to extend the cultural theory of the Douglas school, I will combine the framework with international regime theory, as well as with several concepts and insights taken from the broad approach that has come to be known as 'the new institutionalism'. This reveals the second aim of my research: to explore with which other theoretical approaches grid-group theory needs to be complemented to be applicable in the study of international relations. Cultural theory is still very much in development across a wide range of disciplines. With this book I hope to make a contribution not only to the 'constructivist turn' in the study of international relations, but also to the further development of cultural theory. It should be clear by now that this is an attempt at *exploring theory*. I will make the theoretical argument that both the cultures set out by grid-group theory, and the institutional contexts in which actors operate, are important in order to understand international processes.

The exploration of theory can only take place in empirical research. In this book, I will examine two cases of cooperation on transboundary environmental issues. These cases consist of the domestic and international efforts to protect the natural environment of the Rhine, as well as the national and international attempts to restore the natural environment of the Great Lakes in North America. My main focus will be on the environmental protection of the Rhine. I will use the efforts to preserve the Great Lakes mostly as a basis for comparison. The time-period of my study will be restricted to the years 1970–98. The third aim I have had in writing this study is to gain and provide insight into the conditions under which private and public organizations contribute to the protection of transboundary ecosystems. In my empirical research I have stumbled upon a number of counter-intuitive insights: firms that voluntarily contribute to the clean-up of a water basin; public officials who reject international environmental regulation on the basis of perceived national interests, while at the same time developing extensive environmental protection policies within their own countries; a venerated international organization that hinders the environmental protection it is supposed to facilitate – to name a few paradoxes.

But I would like to emphasize that, to me, the first of the three aims mentioned above is the most important one: to advocate the use of grid-group theory in the study of IR. The extensive case studies presented in the book illustrate how the approach could be applied. The empirical findings are meant to underline the usefulness of the theoretical framework. Too often in the field of world politics, general theories or typologies have been formulated that can only be applied in empirical research with great difficulty, if at all. By including extensive case studies, I hope to show the applicability of grid-group theory in empirical studies of transboundary issues. In the end, of course, the development of theory and the undertaking of empirical research should go hand in hand.

Overview of the book

The arguments of this publication will be presented in the following order. The book consists of three parts, besides the introduction (the present chapter) and conclusion (Chapter 7). The first part concerns theory, and encompasses Chapters 2 and 3. In Chapter 2 I will present grid-group theory, discuss some of the criticisms that the framework has attracted, and defend its application in the study of international relations on theoretical and meta-theoretical grounds. I will argue that the cultural theory of the Douglas school meets a number of criticisms that have been levelled by constructivists at existing (international) political theories. In addition, I will show that grid-group theory suffers less from a few conceptual and methodological problems that have been plaguing constructivist analyses themselves. In Chapter 3, I then extend grid-group theory to the analysis of international regimes and institutions. In order to be able to do this, I will combine grid-group analysis with regime theory and a new institutionalist approach. I will also clarify my use of the concepts 'institutions' and 'regimes'. Chapter 3 will end by introducing three propositions regarding international policy change, the emergence of cooperation in international politics, and the effects of domestic and international institutions on international processes. The first and second of these propositions concern the effects that cultures have on international regimes. Together, they therefore constitute the claim that 'cultures matter in transboundary relations'. The third proposition relates to the impact that domestic and international institutions have on efforts to solve transboundary issues. By itself it forms the complementary claim that 'institutions matter as well in transboundary relations'.

In Part II of this study (entitled *'Cultures Matter'*) I will further explore the first two propositions in empirical analyses of the efforts to protect the Rhine. This part is made up by Chapters 4 and 5. Each chapter starts with an empirical puzzle that is explained on the basis of one of the hypotheses. In Chapter 4, I will explain how in 1987 the intergovernmental cooperation on the environmental protection of the Rhine basin could suddenly change from limited, ineffective and counterproductive to extensive and exemplary. In this chapter I will also argue that cultural theory's conceptualization of policy change needs to be complemented with ideas expressed by Peter Hall, Paul Sabatier and Oran Young. In Chapter 5, I will attempt to explain why industrial firms in the Rhine watershed have made large voluntary investments in water protection, while the agricultural sector and international policy-makers for a long time made no contributions to the restoration of the river basin.

The same procedure is followed in Part III (named *'Institutions Matter as Well'*). In this section, which consists of Chapter 6, I will examine the third proposition, the one on the impact of institutions on international processes. To do so, I will compare the domestic and international efforts to reduce industrial discharges into the Rhine with the domestic and international attempts to decrease industrial effluents into the Great Lakes. The puzzle of this chapter can be formulated thus: how is it possible that the discharges by US firms into the Great Lakes have been more toxic than the industrial effluents into the Rhine, despite the existence of many factors that would have led one to expect the opposite? I will attempt to unravel this puzzle on the basis of differences in domestic and international institutions that shape the relations between actors in both areas.

In the concluding Chapter 7 I will suggest an answer to the question: when do institutions matter more than cultures in international relations?

The logic and limits of this research

At least two familiar arguments can be raised against my research design. First, it can be argued that hypotheses about international processes cannot be properly tested in an analysis of just two examples of international cooperation and strife. In other words, it can be claimed that the use of only two cases strictly limits the strength and certainty of the inferences that can be made. Another objection could be made against the employment of a cultural approach, i.e. against a theoretical framework that focuses on shared perceptions, meanings, norms and preferences. It is sometimes asserted that such concepts are difficult to observe, and that their use in causal reasoning should therefore be kept to a minimum. With some reservations I accept the first objection. However, I do not subscribe to the second.

Using only two cases of international institutions, in my view, certainly limits the extent to which the findings of this study can be generalized to other instances of international cooperation. In a 'probabilistic' world, a sample of two cases is too limited to either strongly confirm or even forcefully reject hypotheses.[13] Therefore, I would like to stress from the beginning that this study is of a tentative character. Its aims are largely theoretical and explorative: extending grid-group analysis to the study of international relations and trying to chart some of the insights that might thus be gained. I will not use the two case studies of international environmental protection as strict tests of cultural theory, merely as opportunities to illustrate and refine the theory. The conclusions that I will draw in my case studies cannot be extended effortlessly to other instances of international environmental cooperation, let alone to other forms of international coordination. My findings will only be suggestive.

However, having said this, I would also like to emphasize that I have incorporated in the research design a number of standard procedures that will increase the strength of my inferences (given the limits imposed by a sample of two cases). Basically, I have tried to follow the advice given by Gary King, Robert Keohane and Sydney Verba to maximize the amount of observations that my cases allow for.[14] The selection of two cases does of course not mean that the number of possible observations equals two. First, in a single case of cooperation on a transboundary ecological issue a variety of actions may be observed. These actions may provide a number of tests of the hypothesis that one wants to look at. Moreover, three comparative methods exist that can be employed to increase observations and to test postulated causal links: cross-time comparisons, the counterfactual approach and cross-spatial comparisons. I have employed these methods in Chapters 4, 5 and 6, respectively.

In cross-time analysis, one examines how the 'values' or 'states' of variables correlate in the same locality over time.[15] By restricting oneself to a single region (where many elements, such as the political and economic system, only change slowly) one rules out a great number of possible explanations of political or social changes. I have followed this procedure in Chapter 4, where I consider not only why an overhaul of the Rhine regime took place in 1987, but also why this regime was *not* fundamentally altered in other years.

The second research strategy, counterfactual reasoning, is of a somewhat different kind.[16] Instead of comparing various real life-events, counterfactual analysis entails comparing a real life-event with an imagined situation. If one wants to link the occurrence of a certain event to a potential causal factor with the help of counterfactual reasoning, one needs to answer the question: would the event still have taken place if that causal factor had had a different value than in reality? This research strategy underpins the conclusions I draw in Chapter 5.

In cross-spatial analysis, undertaken in Chapter 6, one compares the links between explanatory and dependent variables in different regions. The validity of this comparison greatly increases when one selects regions in which the values of all other imaginable explanatory factors are the same (are 'held constant').[17] In Chapter 6, I will follow this research strategy quite meticulously. In this chapter I will compare the toxicity of industrial releases into the Great Lakes with those into the Rhine river. These two areas, the Great Lakes basin and the Rhine watershed, have shared many characteristics. They have had comparable economic structures and functions. The countries in which both ecosystems are located have all been democracies and have been at peace with each other since at least the Second World War. The people populating both water basins have enjoyed comparable levels of wealth. The similarities between these factors mean that these factors cannot explain the differences in the toxicity of industrial releases that I found. In addition to these similarities, a number of (non-institutional) differences between the two cases would lead one to expect the opposite of what I found. This means that only institutional differences between the two areas can account well for my finding that the industrial releases into the Great Lakes have been much more toxic than the effluents of firms into the Rhine.

In sum, in my research I employ three different qualitative research strategies – one for each empirical chapter. Within the limits of case study research, this strengthens the inferences that I draw in this book.

I reject another well-known objection that can be raised against this research. This is the claim that cultural approaches should not be used in causal analysis as they focus too much on shared meanings, perceptions, norms and preferences which are supposedly difficult to observe.[18] I have three problems with this claim. First, I believe that meanings, perceptions, norms and preferences are too important in social life to be ignored, even if it were true that these concepts are difficult to observe and measure. Moreover, I think that this claim understates the problems that are involved in observing other phenomena: nor can concepts such as the state, the international system or the national product be seen walking down the street. The behaviour of people can be observed more directly. However, it is not so much the physical movements of people that we are interested in, but the meanings we and they attach to these movements. It could therefore be stated that similar problems of observation plague the analysis of 'non-mental' factors.[19] Last, Albert Yee has convincingly argued that causal reasoning within political science *necessarily* involves a focus on meanings, perceptions and discourse. According to him, a true casual explanation not only establishes the statistical correlation between social phenomena, but also spells out the motivations of actors involved in these phenomena. Spelling out the motivations of actors inevitably entails an account in terms of meanings, perceptions, norms and preferences – the stuff that cultures are made of.[20]

The nuts and bolts of this research

This research has tried to follow Alexander George's method of structured, focused comparison.[21] The adjective 'structured' here stands for the injunction to apply the same propositions in all cases of the research,[22] and to confront interviewees in all cases with the same set of questions. 'Focused' refers to the need to reconstruct the steps of the policy processes through detailed examination of the actions and motivations of all the concerned people and organizations. In the course of this research, I held 101 interviews with 109 government officials and representatives from international organizations, environmental groups, agricultural associations and companies involved in the pollution and protection of the Rhine and the Great Lakes.[23] These persons were asked to react to (*grosso modo*) the same set of questions. The interviews were held to tap the underlying cultural beliefs of the actors, as well as to reconstruct policy processes. The same objectives motivated examination of governmental files, the reading of materials published by the involved organizations and the consultation of other scholarly work on the environmental protection of the Rhine and the Great Lakes.

I would like to acknowledge the informal manner in which I have processed the data that I obtained during interviews. Ideally, I would have transcribed each of the 101 interviews I have held and then subjected the transcripts to a formal content analysis. Since I am undertaking a cultural analysis, such a strict procedure seems especially called for. However, in my estimate, transcribing all interviews would have taken me about nine months. And then I would still have faced the task of conducting a formal content analysis. Instead I have opted for the following, more informal, procedure. After holding the interviews, I relistened to all of them on tape. While doing so, I made notes on the cultural dispositions of the groups of organizations that I wanted to analyse. This gave me a clear overview of the ways of life to which the main actors adhere. In the writing stage I double-checked my assertions by again consulting the cassettes. Whenever I made a specific assertion about the cultural tendency of a group of actors, I listened again to the taped interviews I held with these organizations. Thus I went through three stages: the interviews themselves, a next round of listening to all of the tapes and a last round of checking my specific assertions by hearing the relevant cassettes. This procedure has taken me some three months to complete. In my estimate, the costs of a formal content analysis (in terms of time) were too high, and the benefits of such an approach (in terms of more reliable information) were too small, to justify taking this route. But I could fully understand if others were to disagree.

A note on terminology

The stampede of calls for new approaches, theories and paradigms in the study of international relations that began some ten years ago has also led to a lot of terminological innovation and confusion. Are we analysing 'international relations', 'transnational relations' or 'world politics'? Or should we rather use James Rosenau's term 'postinternational politics'?[24] Is international political economy a subfield of the study of world politics or the other way around? Personally I prefer the term 'transboundary relations'. The terms 'international' and 'transnational' literally refer to links between *nations*, which is only a small part of what we study. The concept 'world politics' too much evokes intergovernmental dealings. The transactions of private actors across borders are hugely important as well. By contrast, 'postinternational politics' seems too vague – it could apply to anything. The term 'transboundary relations' is clear, neutral and encompasses all those practices of private and public actors that we study. But I will not be very rigid. I will often use the denominator 'international relations' when strictly speaking I mean 'transboundary relations'. And I will continue to speak of 'the study of international relations' and 'the study of world politics' when, properly speaking, I think I should be talking of 'the study of transboundary relations'.

One common conceptual mistake I will not make. I will not conflate state and society. When I write of 'the state' I will mean government personnel and departments, and not other parts of society. So many political debates and struggles are waged over the strength and size of the state that I find it unacceptable to merge the concepts of government and society. In constructing domestic state–society relations, actors also construct a certain international system (and vice versa).[25]

Part I
Theory

2
Grid-Group Theory and the Study of International Relations

This chapter will advocate the application of grid-group theory in the study of international relations on (meta-)theoretical grounds. The main argument will be that grid-group theory meets several conditions for proper theorizing that have been formulated by constructivist scholars, while also offering a way to avoid some problems that have plagued constructivist empirical analyses. First, the chapter will briefly describe the constructivist approach to international relations. Thereafter, grid-group theory itself will be set out and some of its conceptual limitations and strengths will be discussed. The chapter will then show how cultural theory not only meets the ontological and theoretical demands made by constructivists, but how it could also be helpful in overcoming various shortcomings of empirical research that has been based on constructivist assumptions. Last, several possible uses of grid-group analysis within (international) political theory will be listed.

The constructivist challenge

As argued in the Introduction, the void created by the demise of the contending approaches in the inter-paradigm debate (neorealism, pluralism, structuralism) has been increasingly filled by the theoretical claims and empirical analyses of a number of scholars who have grouped themselves under the banner of constructivism.[1] Although various frameworks go under this flag, it can be argued that constructivist approaches share several assumptions. First, constructivists give pride of place to collectively shared norms, values and perceptions. The decisions and deeds of international actors are assumed to be directed by their identities (which can be defined as sets of collectively shared norms, values and beliefs). The material world matters, of course, but it only matters to the extent that collectively held norms, values and perceptions tell people that it matters (and how it matters). An example may be helpful. Both Britain and India possess nuclear weapons. In the case of Britain, this is not at all perceived as a significant problem by the countries in its vicinity, Belgium for instance. For India the situation is quite different. Its ability to launch nuclear weapons is viewed as a grave threat by its neighbours, especially Pakistan, but also China. The same material condition (the possession of nuclear weapons by a government) has a quite divergent impact in different regions, as the collectively held identities of the people living within these areas vary greatly. The citizens of India and Pakistan have come to define their identities in opposition to each other, while the populations of Britain and Belgium no longer do so. Many other approaches within the social sciences postulate that people are driven by only a few, simple motives, for instance lust for power or money. As these approaches assume that all actors are motivated by uniform concerns, the motivations and perceptions of actors cannot explain why certain social processes take place in particular times and places and not in other times and places. As a consequence, in these frameworks differences in material conditions (e.g. birth rates, weapon arsenals, wealth) become the only explanatory variables that can account for differences. Constructivists typically allow for a richer variety of motivations, including divergent perceptions, norms, values and identities. In their empirical research, identities can therefore play a larger explanatory role.

Second, constructivists argue that the social structures in which we live are upheld by our identities. In the constructivist view, people 'act out' their norms, values and beliefs. Through their words and behaviour they show their allegiance to certain norms, values and perceptions. In doing so, they collectively construct and sustain the structures of their social (as well as physical) world. In other words, in constructivist thinking social relations are not (primarily) material, but are made up of collectively held norms, values and meanings. Consider for instance, a social environment that is hostile to people with a homosexual orientation. Such a social structure may be underpinned by formal rules, such as the legal right to expel homosexual people from the teaching profession or the army. Yet these formal rules would not be very effective if they were not underpinned by a negative attitude among people.

In a social environment that is hostile to homosexuals, people will frequently make jokes about gays, stay away from them, talk about them in negative ways. All these words, gestures, ideas and emotions make up the hostile social environment. In fact, the discriminatory formal rules themselves should properly be seen as a part of this social fabric. They do not only have a regulative impact (in providing a legal basis for excluding gays from several sectors of the labour market), but a symbolic one too that sustains the social order. This view has important implications for how social change is understood. If social structures are upheld by people's norms, values and beliefs, then people can also transform these social structures by changing their predispositions. This will not, of course, be an easy task. Norms, values and beliefs are typically *collective* entities. A single person does not create a norm or value by him- or herself. Changing identities, and the social structures they support, will always be a collective undertaking. Yet, on occasion an individual (or group of individuals) may be quite effective in inducing other people to adopt new norms, values and beliefs.

Third, constructivists are not 'anti-foundationalist'. This term is a favourite among poststructuralist IR scholars, who make up the other school of thought that has become prominent in the study of world politics since the late 1980s.[2] It signifies the belief that valid generalizations about social life across time and space are well-nigh impossible to make. This belief rests on the assumption that language is recursive. People cannot express themselves but through language. Therefore, according to poststructuralists, words uttered or thought by people do not so much refer to objects, but to other words, which in turn refer to yet other words, and so on and so forth. As a result, words/ideas/opinions do not reflect so much objects in 'reality', but reflect other words/ideas/opinions.[3] This means that every attempt at formulating an objective hypothesis that may be valid across time and space is doomed from the beginning, and can only 're-present' a picture of the world that is coloured by highly specific personal and cultural assumptions. It does not say a great deal about the 'real world out there', only something about the views of the observer.[4] Generally speaking, constructivists do not share these beliefs. They seek to find a middle path between a very simplistic correspondence view in which words directly represent objects in reality, and an equally simplistic one according to which words do not describe reality at all. They acknowledge that scholars and other observers of social processes always bring their own cultural biases to their research. But by being aware of this (by being 'self-reflexive'), as well as by strictly following the rules for scientific inference, they may still be able to find empirical generalizations that are valid across time and space and that can be collectively agreed upon.

Constructivists have formulated at least four (mainly ontological) conditions that proper IR theories should meet. They have argued that the theoretical approaches that were dominant in the study of international relations until the late 1980s failed to meet these conditions, and as a consequence presented an impoverished view of international processes. I will present these four requirements below. In the rest of the chapter I will attempt to argue that grid-group theory satisfies these four demands, and can therefore be seen as a constructivist approach.

First, IR theories should describe the collectively shared norms, values and perceptions of actors. These predispositions compose the social structures that confront people. Second, IR approaches should capture the ways in which actors and social structures mutually constitute each other.[5] As mentioned above, actors sustain the social structures in which they live by acting out their norms, values and beliefs. As such, actors have a degree of freedom in transforming their social relations by changing their collectively held views. But social structures also shape individual actors. People partly adopt their norms, values and beliefs from the social structures in which they find themselves. IR theories should attempt to chart the ways in which actors and structures shape each other. Third, IR analyses should be able to explain fundamental international change.[6] Last, IR analyses should be undertaken in a self-reflexive manner.[7] Below, I will show that grid-group theory meets these conditions – or at least does so more fully than many other IR approaches have done in the past.

Grid-group theory

In 1990 a new cultural perspective on politics and society was boldly presented with the publication of *Cultural Theory*.[8] In this book, Michael Thompson, Richard Ellis and Aaron Wildavsky clarified and systematized seminal work by Mary Douglas, whose aim it had been to apply anthropological methods and insights to contemporary Western societies.[9] Rather immodestly, they named their analysis 'cultural theory', although it has also become known as 'grid-group theory' and as 'the cultural analysis of the Douglas school'.[10] Outside the study of international relations, cultural theory has become the topic of a lively debate between proponents and critics of the approach.[11] The approach has already been employed in the fields of anthropology,[12] sociology,[13] political science,[14] public administration,[15] the history of science and technology,[16] risk analysis,[17] as well as the study of religions.[18] By contrast, grid-group theorists and students of international relations have, on the whole, ignored each other. Grid-group theorists have paid scant attention to transboundary processes, while very few international relations scholars are familiar with (this brand of) cultural theory.[19] This book is an attempt to interest academics in bridging this gap.

Grid and group: building blocks of cultural theory

According to the cultural theory of the Douglas school, people are *moral actors*. That is to say, people care deeply for how their social relations are organized. Actors are supposed to have strong preferences regarding the amount of solidarity within society, the extent of personal freedom, and the ways in which authority relations are structured, to name but a few moral issues that people are concerned with. Furthermore, grid-group theory assumes that people feel compelled to justify their moral views – not only to others, but also to themselves. As part of this, they seek to discredit other moral positions, i.e. to portray as immoral alternative views on how social life should be organized. People attempt to achieve these goals by making a host of claims about empirical reality that make it look only 'natural' or 'self-evident' that their preferred way of living is the 'right' way. For instance, if some aspects of the economy are not organized in accordance with the (moral) preferences of a group of people, then these persons will claim that this situation will have various undesirable effects, such as rising unemployment, declining wealth, a polluted environment and higher crime rates. They will claim that these dire consequences can only be averted by bringing economic relations more in line with their favoured moral views.

Cultural theory postulates that people are especially concerned with two aspects of social life: group and grid.[20] The *group dimension* represents the extent to which people are restricted in thought and action by their commitment to a social unit larger than the individual. High group strength results when people devote a lot of their available time to interacting with other members of their unit.

In general, the more things they do together, and the longer they spend doing them, the higher the group strength. Where admission to the social unit is hard to obtain, making the unit more exclusive and conscious of its boundary, the group strength also tends to be high. An extreme case of high group strength is the monastic or communal setting where private property is renounced upon entering and the members depend on the corporate body for all of their material and social needs. High group strength of this sort requires a long-term commitment and a tight identification of members with one another as a corporate identity. Individuals are expected to act on behalf of the collective whole, and the corporate body is expected to act in the normative interests of its members. Group strength is low when people negotiate their way through life on their own behalf as individuals, neither constrained by, nor reliant upon, a single group of others. Instead, low group people interact as individuals with other individuals, picking and choosing with whom they will associate, as their present preoccupations and perceived interests demand. The low group experience is a competitive, entrepreneurial way of life where the individual is not strongly constrained by duty to other persons. Attractive though this freedom from constraint might first appear to some of us, there is a serious disadvantage: the low group individualist is unable to fall back on the support of her fellows should her personal fortune wane. In the high group context, the safety net of social support is compensation for the loss of personal autonomy.

The *grid dimension* of social life is the complementary bundle of constraints on social interaction, a composite index of the extent to which people's behaviour is constrained by role differentiation, whether within or without a group. Grid is high strength whenever roles are distributed on the basis of explicit public social classifications, such as sex, colour, position in a hierarchy, holding a bureaucratic office, descent in a senior clan or lineage, or point of progression through an age-grade system. It is low strength when classificatory distinctions only weakly limit the range of social choices or activities open to people. A low grid social environment is one in which access to roles depends on personal abilities to compete or negotiate for them, or even of formal regulations for taking equal turns. In either case, access to roles is not dependent on any ascribed characteristics of rank or birth. In other words, where roles are primarily *ascribed* grid constraints are high. Where roles are primarily *achieved* grid constraints are low.

In grid-group analysis it is assumed that many of the perceptions, values and opinions that people have are related to their
preference for a specific combination of the grid and group dimensions. It is expected that when a number of conflicting opinions
and values exist, people will tend to adopt the beliefs and norms
that will be most compatible with their preferences for a certain
combination of grid and group. This is the case because of two
reasons. First, it is because cultural theory assumes that people are
inherently moral actors who are continuously engaged in justifying themselves to themselves as well as to others. In the process of
justifying themselves, actors make all kinds of empirical claims that
are meant to convince everyone of the rightness of their actions
and views. Second, it is because cultural theory assumes that the
grid and group dimensions capture those elements of social organization that people care most for.[21] In grid-group theory, the values
and opinions that people have are by and large consistent with
each other, as they tend to be based on the same ultimate preference for a certain way of structuring social life. These coherent sets
of values and opinions are called *cultural biases*.

Assigning two values (high and low) to the grid and group dimensions gives four possible cultural biases, justifying four ways of
organizing social relations. The combination of a high score on
the grid dimension (preference for many rules that prescribe people's
roles) with a high score on the group dimension (preference for
strong group boundaries) gives the *hierarchical* cultural bias. A high
grid score (preference for many rules that prescribe roles) with a
low group score (preference for weak group boundaries) characterizes the *fatalistic* cultural bias. *Individualism*, the third cultural bias,
is associated with low scores on both the grid scale (dislike of many
prescriptions) and on the group scale (in favour of weak group limits).
Last, *egalitarianism* is the cultural bias that is associated with a low
grid count (dislike of many fixed roles) and a high group count
(preference for strong group boundaries).[22] In cultural theory it is
argued that the beliefs, perceptions and values of persons fall into
one of these four categories.

Figure 2.1 illustrates the points made above. Two examples will
perhaps help to illuminate these cultural biases. The first example
concerns people's perceptions of human nature; the second is about
people's views of nature.

There is much disagreement about how people *really* are. Some
of us tend to believe that human beings are essentially good-natured;

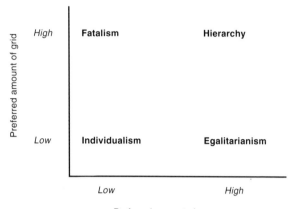

Figure 2.1   Cultural biases

others have a much more cynical outlook, and tend to assume that people are purely driven by greed and lust after power and status; and others again claim that humans are completely unpredictable. Grid-group analysis tries to explain why different individuals come to adhere to these various beliefs.[23] Egalitarians believe that humans are intrinsically good but are corrupted by evil institutions. They need to see human beings as intrinsically good, because otherwise they could not justify (to themselves and others) a life of voluntary association and solidarity. By arguing that all the evil, greed and maliciousness of this world is generated by coercive institutions and regulations, egalitarians are able to provide a defence for their belief that people are intrinsically good in spite of all the grave social problems in the world. Moreover, by attributing all the evil of this world to coercive institutions, they also attack the social orders that are favoured by hierarchists and fatalists.

Individualists believe that humans are incurably self-centred. Institutions that try to change or rein in people's concerns with their own happiness and wellbeing can therefore not be successful. Moreover, individualists will argue that such institutions are not necessary. They believe that a 'hidden hand' operates, not only in the marketplace but also in other parts of social life, which ensures that attempts to satisfy one's own desires also benefit other people.

Hierarchists justify their own preferred form of social order by assuming that human beings are born bad but can be redeemed by good institutions. If social life were left unregulated, then society would be a war of all against all and life would be brutal, nasty and short. However, if the right prescriptions and procedures are instilled in society, then people can be saved from themselves.

According to fatalists, human nature is unpredictable, and it is therefore impossible to know whether to trust or distrust someone. It therefore appears rational and safe to fatalists not to trust anyone easily, as this will avoid putting confidence in people who cannot be relied upon. This cautious and sceptical attitude sustains and justifies the social order that fatalists have come to accept and prefer. If no one is to be trusted, then it will seem useless, and even undesirable, to try to cooperate with other people in an attempt to free oneself of the externally imposed regulations that prescribe one's life.

A second illustration of the cultural biases distinguished in grid-group theory concerns alternative myths of nature. People not only argue about how human nature 'really' is, but also about the extent to which natural systems can accommodate human activities. Some believe that ecosystems cannot be affected much by human production and consumption. Others constantly fear the imminent collapse of nature. Again, cultural theory enables one to sort out which groups of people feel inclined to which particular belief.[24] Individualists believe in the myth of *nature benign*. The world, this myth tells us, is wonderfully forgiving. No matter how people behave ecosystems will not be affected much. Natural systems are quite resilient and stable. This view of nature encourages and justifies the individualistic preference for total, personal freedom. No government institutions are needed to rein people in with a view to protecting the environment.

Egalitarians will espouse the myth of *nature ephemeral*. According to this myth of nature, the world is a terrifyingly unforgiving place. Even the smallest amount of human activity may bring about ecosystem collapse. Natural systems should be treated with great care. This view of nature legitimizes the egalitarian way of organizing social relations. It does not leave room for extensive efforts to produce and consume – major sources of inequality. Instead, it justifies living in small, tight-knit, decentralized communities that respect nature's fragility and make appropriately modest demands upon it.

Hierarchists feel comfortable with the myth of *nature perverse/ tolerant*. In this view, ecosystems are usually quite resilient. As a rule, ongoing human activities will not affect natural systems too much. Sometimes, however, they do. If consumption and production patterns transgress certain boundaries ('get out of hand'), ecosystems will collapse. It is up to the experts and authorities to discover where these boundaries are, and to make sure that everyone stays within the boundaries. This myth of nature therefore endorses central planning and regulation, which are essential elements of the way in which hierarchists want to organize social life.

Fatalists embrace the myth of *nature capricious*. This myth portrays a random world. According to this belief, it is not possible to learn or manage ecosystems. The impact that human activities have on the physical world is not predictable. All that can be done is to cope with erratic, unforeseeable events.

Grid-group theory distinguishes many more issues about which people form opinions that justify their own preferred form of social organization. These issues include ways of matching needs and resources, ways of assigning blame when things go wrong and attitudes towards risk. Moreover, one can often make predictions about people's perceptions and beliefs with regard to specific policy issues on the basis of the social order that they claim to prefer and their associated views of nature, human nature, risk attitude, etc.[25] It is of paramount importance for understanding grid-group theory to grasp the 'trick' involved in the process of linking specific views and norms to any cultural bias. Without understanding this trick grid-group theory could not be applied. The trick consists of asking oneself: if someone were to adopt a certain view, perception or norm, would it become easier or more difficult for that person to justify a certain pattern of social organization. For instance, someone advocating a realignment of social conditions along more egalitarian lines would find it much harder to be convincing if ecosystems were portrayed as highly resilient. If nature was seen as highly resilient, there would seem to be less reason to keep people's consumption and production patterns (important causes of inequality) in check. In cultural theory, particular norms, beliefs and values are related to the various cultures on the basis of their effects on people's preferences for certain forms of social order.

Thus far, the cultural biases set out by grid-group theorists have seldom included perceptions of, and preferences for, international issues. But there is nothing in the theory that would stop one from doing this. In Chapter 3, I will extend the four cultural biases of grid-group analysis to the domain of international relations. I will show that adherents to the different cultural biases have quite divergent perceptions of, and preferences for, a range of transboundary issues. One of the main strengths of grid-group theory is therefore that it can provide a systematic and coherent taxonomy of the ways in which actors perceive and construct all kinds of international issues. Such a taxonomy is important for empirical research as (a) it might enable researchers to condense the multitude of ideas, values and perceptions, and wishes of actors into a limited, manageable set of rationalities; and (b) it would make different cases comparable. Furthermore, when applied in empirical research on international issues, I believe that grid-group analysis would be able to provide testable explanations, as the theory would link the specific policy views of the involved international actors to their underlying political preferences, views of (human) nature, risk attitudes and ways of assigning blame. Part of the explanatory power of grid-group theory is therefore based on the connections that the theory makes between specific policy beliefs by people and their underlying cultural biases which tell them how to go about realizing their policy preferences.

Four ways of life

Grid-group theory not only postulates that perceptions and norms are related to preferences for certain forms of social order. The theory also illuminates aspects of how people come to prefer certain ways of organizing social life, and their concomitant beliefs and values. It is argued that the social structures in which individuals actually live limit the range of perceptions, preferences and values that become available to these individuals. Social structures serve as a kind of filter for human cognition. Grid-group analysis assumes that social structures tend to filter out those beliefs, perceptions and values that do not justify and thereby sustain the social structures themselves. The theory therefore postulates that groups of people tend to adhere to those preferences, perceptions and values that sustain the social settings in which they live.

This should not be seen in a deterministic way. In cultural theory, individuals are portrayed as rational, creative agents. They think for themselves. Sometimes, people's preferences and perceptions will contain elements that do not legitimate the social setting in which they live. Also, people may grow disenchanted with their social environment and the kind of ideas that it allows for, and develop quite different beliefs. However, grid-group theorists expect that in adopting a new set of beliefs, these people will also switch (or be expelled) to a social setting that is more compatible with their new beliefs. Alternatively, they may seek to change the social setting in which they live itself – in order to make it more compatible with their new views. By and large, therefore, cultural theory predicts that the persons who find themselves in a certain social setting have a tendency to adhere to the perceptions, preferences and values that sustain their social relations.[26] More specifically, it is expected that people who find themselves in an egalitarian social context (i.e. social relations that are characterized by few regulations and strong group ties) will have the tendency to adopt many of the values, perceptions and preferences that belong to egalitarian bias. Persons who live in an individualistic social structure (with neither many rules nor strong group ties) tend to adhere to many elements from the individualistic cultural bias, etc. Together, cultural biases and their concomitant social structures are called *ways of life*.[27]

Several points of clarification may be useful. First, the distinction between cultural biases on the one hand, and social relations on the other, does *not* correspond to the distinction between 'material/ observable' and 'mental/unobservable' factors. A cultural bias encompasses all those norms, values, perceptions and beliefs that legitimize a certain way of organizing social relations. Cultural biases therefore contain only mental elements. But the sets of social relations distinguished by cultural theory do not consist solely of material and directly observable factors. These sets of social relations are made up of the group and grid dimensions. Group stands for the extent to which people are restricted in thought and action by their commitment to a social unit larger than the individual. Grid captures the degree to which people's thought and behaviour is constrained by status differentiation. Both social dimensions contain mental as well as material elements.

Second, it is important to note that the ways of life and cultural biases of grid-group theory should be seen and used as (Weberian) *ideal types*. The reference, above, to 'fatalists' or 'individualists', etc., was a reference to fictional characters whose thought and behaviour fully coincided with only one way of life. In reality of course, few actors will fully adhere to a single culture. Normally, the thought and behaviour of actors will include elements from various ways of life, and will also probably vary from one social domain to the other. However, both the way of life and the cultural bias concept will only be useful as explanatory variables if the ideas and deeds of single actors within a specific social domain in the main fall into one cultural category. Whether this is indeed the case is an empirical question, not a theoretical one. In addition, it should be realized that, as ideal types, ways of life and cultural biases will differ somewhat from time to time and place to place.

Third, cultural theory is applicable at multiple levels of analysis. The ways of life concept can be used for any conceivable group of people – from the individual household all the way up to the international system. One can for instance talk of individualistic, fatalistic, hierarchical and egalitarian households, but also of international systems that are individualistic, fatalistic, egalitarian or hierarchical. In other words, the theory sees the cultural battles between the four ways of life as being played out at all levels of analysis.[28] In the empirical part of this study cultures will mostly be distinguished at the organizational level (firms, government agencies, citizens' groups).

In cultural theory, political and social life is viewed as a never-ending struggle between the hierarchical, individualistic and egalitarian ways of life. The proponents of these ways of life constantly try to impose their rationality on each other and to institutionalize their preferred form of social order. The fatalists are often on the receiving end of all this activity. They are usually not politically active or organized and, most of the time, accept whatever is decided for them by the proponents of other ways of life.

From the viewpoint of grid-group theory, international politics is created by adherents to the four ways of life who are all trying to impose their interpretations on others and who are all trying to institutionalize their preferred form of social relations. These actors work for the various organizations that make up the world system, including government agencies involved in any aspect of foreign policy, multinationals, international organizations, human rights organizations, environmental groups, the media and universities. Often, adherents to different cultural biases have to use their power resources to fulfil their wishes, or else they have to reach political compromises or come up with creative policy solutions that satisfy the wishes of the adherents of all ways of life.

The normative aim of grid-group theory is to increase the tolerance of, and mutual respect between, different ways of acting and thinking. The theory postulates that each of the four ways of life has specific and important contributions to make to the public debate, while each way of life also has its own shortcomings and blind spots. It is predicted that open political systems (i.e. political systems in which all organizations have full and unrestrained access to the decision-making processes) will be the most viable political systems.[29]

So what about those two old favourites of social explanation: the drive to satisfy one's interests and the yearning for power? Regarding the latter it must be observed that an emphasis on power struggles is fully compatible with grid-group theory. The framework assumes that actors attempt to impose their preferred morality by deploying all kinds of power resources, ranging from gentle persuasion to the throwing of bombs. However, on the basis of the approach it can also be postulated that the various cultures include alternative views on which means of power are legitimate and effective and which are not, and also on the goals for which the means of power should be used. But in principle there is no contradiction between cultural theory and the view that an important part of social reality is made up of power struggles. With regard to interests, it must be noted that this is largely a vacuous concept. Whose interests? Which interests? Those are the important, and often unanswered, questions. Grid-group theory is actually one of the few approaches that spell out the alternative interests that people perceive. For instance, it postulates that the behaviour assumed by standard neoclassical economic theory (people constantly seeking to expand their financial resources in order to satisfy their never-ending needs and desires) occurs most often in individualistic spheres. In a hierarchical setting, material gain is condoned and encouraged as long as the stratified social order is not at risk. Cheating as a way to prosper (or survive) abounds more under fatalistic conditions, while any form of enriching oneself is usually frowned upon in egalitarian circles.

Theorizing social change

Grid-group analysis offers a dynamic account of social and political life. It highlights three ways in which social and political change can take place.[30] First, cultural theory sets out how fundamental, large-scale social transformation may occur. The theory assumes that the hierarchical, individualistic, fatalistic and egalitarian ways of life are always present in any major issue-area and within society as a whole. However, the number of people that will adhere to each of these ways of life is expected to wax and wane over time. Each large defection from one cultural corner to another sets off fundamental social change. This transfer of loyalty from one way of life to another is expected to take place as follows. When the adherents to a particular way of life have been dominant in the political process for an extended time period, the particular blind spots and weaknesses of their cultural bias will lead to unforeseen social problems and tensions. For instance, individualists view the natural environment as 'benign' and bountiful. According to this view, ecosystems do not easily collapse under the pressure of human activities, and everyone should be allowed to invest, produce and consume as much as they like. However, it may turn out that the actual ecosystems in which people live are not as resilient as the individualistic bias suggests. In that case, a society in which the individualistic way of life has held sway for an extended time period, will, in the end, experience a number of unexpected ecological problems. When within a political system the dominance of a way of life has led to the accumulation of a number of unforeseen social problems, more and more people (being active and rational) will start to switch to ways of life that seem to capture the world more accurately. As a result, these other ways of life will become more dominant in the political system, and their political representatives will be able to acquire more power resources with which they can restructure society according to their wishes. This account of fundamental social change is known within grid-group analysis as the 'theory of surprises'.

The second route to social change that is stipulated by cultural theory concerns more small-scale changes. In responding to changes in circumstances, adherents of a culture first engage in 'problem-solving' efforts. They are more likely to adapt to changing conditions by adjusting their policies and arrangements so as to better support their way of life rather than transfer their allegiance to a rival culture. The latter will only take place after surprises have occurred that cannot possibly be explained or accommodated by their ways of life. Before this, however, all kinds of small-scale, stopgap measures will be taken that are compatible with the larger concerns of each culture, the preferences for grid and group.

The third kind of social and political change that is highlighted in cultural theory consists of shifting political coalitions. In cultural theory, political parties and governments are often portrayed as coalitions between adherents to different cultures. The approach also points out which attributes a way of life has in common with other ways of life (for instance, in the case of individualism and egalitarianism, a rejection of extensive stratification), and which elements it does not share with other ways of life (e.g. egalitarianism includes support for group solidarity, individualism does not). The former set of attributes offers adherents to different ways of life a basis for forming political coalitions among each other. The latter set of characteristics tends to unravel parties and governments based on a coalition of different ways of life. Therefore, an additional way in which grid-group theory can contribute to the explanation of political change is by identifying the attractors and repellents that underlie shifting coalitions among cultures.

Agents and structures

Grid-group theory provides a possible, partial answer to the agent-structure puzzle in social theory. This puzzle stems from two commonsense beliefs: (a) the behaviour and ideas of individual actors is shaped and constrained by the social structures and institutions in which they operate; (b) social structures and institutions are (partly intentional) outcomes of actions undertaken, and thoughts thought, by actors. The agent-structure problem consists of constructing a theoretical framework that captures this mutual interdependence of individual actors and social structures. As has been pointed out above, according to grid-group theory, social structures do constrain and shape individual behaviour and thought. The social relations in which people live give them their cultural biases and thus serve as a basis for action. However, the cultural biases of all individual actors together determine the social structures in which actors live. Plus, grid-group theory portrays actors as rational and inventive. They are actively involved in assessing the degree to which their personal experiences coincide with the expectations generated by the four cultural biases, and they will switch from one bias to another if they feel that this latter bias provides a more adequate interpretation of daily life. Thus, cultural theory tries to capture important elements of the mutual relationships between actors and structures.

Conceptual challenges and problems

Grid-group theory is certainly not without its own flaws and problems and is still very much in development. The theory has attracted a number of criticisms. Some of these criticisms challenge the basic utility of the project, while others call for adaptations of the theory. A common, all-out attack on cultural theory is the lament that the theory is based on 'just' four categories. This argument would claim that grid-group theory, by suggesting that there are only four basic types of thinking and acting, overlooks the great variety of alternative ways of living and perceiving. It is asserted that, although grid-group analysis uses the terminology of a cultural approach, it disregards the multitude of cultural expressions that are available to people.[31] Advocates of cultural theory have taken issue with this argument. First, they have pointed out that many of the great social scientists of the past used a smaller number of ideal types in their work. Max Weber, for instance, distinguished between only three styles of authority: charismatic, legal and traditional domination.[32] These theorists claim that one of the strengths is that it actually *adds* a cultural bias that has hitherto been neglected in the canons of social theory, namely fatalism.[33] Second, grid-group theorists have argued that they have limited themselves to four categories in an attempt to explain as much as possible with the help of as few variables as possible.[34]

A related, slightly different, criticism is that grid-group analysis is an 'ahistorical' approach: the theory does not allow for the impact of specific historical circumstances on the actions and ideas of people.[35] Rejecting this claim, Richard Ellis has argued that theory and history should not be seen as inimical, and that history provides the testing ground for the propositions of cultural theory. Moreover, the four ways of life of grid-group theory should be seen as ideal types. They are meant to capture basic elements of the ways in which people perceive and act. The precise content of these ways of life are partly determined by the actual historical circumstances in which people find themselves.[36]

Some critics of cultural theory have asked where the grid and group dimensions come from. These dimensions are the central explanatory variables in the approach, but are not explained themselves.[37] In response, Aaron Wildavsky has restated cultural theory's aim of using the smallest number of variables to give the largest pay-off in explanatory power. This aim, he argued, provides the main rationale for choosing the grid and group concepts.[38]

Another conceptual problem concerns the level of analysis at which ways of life should be distinguished in empirical research. Grid-group theory states that the four cultures can be found at any level of analysis. This is an interesting feature of the approach, but in empirical applications it creates the problem at which level of analysis ways of life should be conceived. To solve this problem, a more 'anthropological' attitude may be required of political scientists. They cannot rely on an a priori idea of the exact level of analysis at which the four ways of life may or may not exist. Instead, they have to let the subjects of their study 'speak for themselves'. By interviewing decision-makers and analysing their writings, researchers have to try to discover the groups in which these people partake and the regulations to which they are subject. In this manner, researchers may be able to establish the level of analysis at which the grid and group dimensions (and the related biases) can best be conceived in the particular issue-areas of their empirical studies. In the research presented here, ways of life will in the main be distinguished at the level of the organizations involved in the policy processes. It will be assessed to what extent government departments, environmental groups, citizens' groups, business firms, water supply companies and agricultural organizations adhere to a certain culture.

An additional problem is the 'multiple selves' puzzle: do people stick to a single way of acting and thinking in all aspects of their lives, or do they adhere to different rationalities? This riddle has also confounded other political theorists.[39] It constitutes a crucial question for grid-group theory. Ways of life could only be used as explanatory factors if: (a) actors followed a single culture in all the domains of their lives; or (b) actors stuck to a single way of life in each specific domain of their lives. If, on the contrary, actors mixed and mingled elements from different ways of life in each distinctive social domain, then the four cultures set out by the Douglas school would lose much of its potential explanatory power. Eero Olli has recently provided some empirical insight into the matter.[40]

In his statistical research on environmentalism in Norway he makes a distinction between the 'coherent individual' (who follows a single, internally coherent cultural bias in all social contexts), the 'sequential individual' (who has different cultural biases for different contexts) and the 'synthetic individual' (who has a single rationality for all contexts, but one that includes elements from alternative ways of life). He finds that 40 per cent of the respondents to two extensive surveys can be classified as coherent individuals, 37 per cent as sequential individuals while 23 per cent are synthetic individuals. These results seem quite hopeful.

A related problem is the falsifiability of the theory. Sometimes it is argued that grid-group theory is not falsifiable, especially when the ways of life set out by the theory are treated as ideal types. I do not share this view. If it turned out that the rationalities to which actors adhere typically are combinations of the four cultures set out by grid-group theory, then the framework would be falsified. This does not mean, however, that a small set of instances of actors who mix ways of life would necessitate abandoning the approach.

At least one attempt to falsify cultural theory has been made.[41]

Another consideration is that in empirical research cultural theory often needs to complemented by other approaches.[42] The theory states that four cultural biases exist in any major policy area. This thesis cannot always provide a full account of how policy decisions are made and implemented. Such an account must also take into consideration the relative strength and power resources of the organizations that represent the cultural biases, the institutional context in which the actors operate, the specific policy issues, the cleverness of individuals, etc. This means that, in empirical research, it is useful to combine grid-group theory with other theoretical frameworks, such as approaches that focus more on power politics and institutional context. In this research, grid-group analysis will be combined with an institutional approach (see Chapter 3), and will also be attentive to the strategic moves that actors make.

To these various criticisms and challenges I could add a few remarks of my own. Grid-group theory derives people's cultures from just two dimensions (their preferences for the amount of rules and the strength of group boundaries). Thus it suggests that it is the mere existence of rules and group boundaries rather than the *content* of these rules that is of paramount importance in politics. Yet the content of existing social rules and group bonds may also be quite important. For instance, in one article Dennis Coyle has labelled as hierarchical both a number of mainstream American environmental thinkers and racist organizations advocating segregation and zoning laws.[43] Clearly, there are immense differences between these sets of ideas, which could be expected to lead to quite divergent actions. This problem is aggravated by the fact that the ways of life are very broad categories: as the example above shows, many phenomena can be labelled as hierarchical, egalitarian, individualistic or fatalistic. Case study research, in particular, may be hampered by this. Empirical analysis on the basis of grid-group theory needs to proceed very carefully.

Grid-group theory and constructivism

In the above discussion of grid-group theory it was shown that the approach meets three of the four standards that have been laid down by constructivist IR scholars. These conditions are: (a) conceptualization of the rationalities to which actors may adhere; (b) a description of how actors and social structures may interact;

and (c) an account of the ways in which social change may take place. The fourth constructivist condition also holds for cultural theory. The approach can be used in a reflexive manner. As it stipulates the categories according to which people think and act, it can also be employed to unveil the cultural biases that have informed academic analysis. Wildavsky's volume *Culture and Social Theory* is an attempt to do just that.[44] In this book he employs cultural theory to uncover the particular cultural biases underlying a number of social and political theories. He also shows how these biases have limited the explanatory power of these frameworks. For instance, in this vein he deconstructs several of the core concepts of economics and rational choice theory, including the notion of externality, the distinction between public and private goods, and the assumption of egotistical, uncooperative behaviour. Cultural theory could of course not only be applied to unveil the biases in other people's work. It could also be applied in a self-critical, a self-reflexive, manner.

It can therefore be concluded that grid-group theory meets the requirements for 'good theory' that have been formulated by constructivist IR scholars. In my view, it also does something else. The approach is less vulnerable to several problems that have plagued constructivist analyses thus far. Two sets of charges have been levelled against the constructivist enterprise: a theoretical one and a methodological one. Below, these sets of concerns will be presented with a discussion of why grid-group theory is less vulnerable to these charges than the constructivist analyses that have thus far been undertaken.

The first set of criticisms concerns the paucity of constructivist *theory*. In the beginning of their enterprise, constructivists restricted themselves to critiques of existing IR approaches and to formulating ontological conditions for proper IR theory. Thereafter, constructivists started to undertake empirical analyses as well.[45] The missing link in this chain has been the development of theoretical propositions. Martha Finnemore, herself a prominent constructivist, has put it thus:

What exactly are the social norms that structure and guide contemporary politics?' . . . Little theorising has been done about this. Constructivism itself only claims that social facts influence behaviour; it makes no substantive claims about what those facts are more than rational choice makes claims about the content of interests. Scholars in the American political science community whose work I would label constructivist have taken one of two directions, neither of which fills this void. Some, like Wendt and Kratochwil, have concentrated on elaborating more abstract social theory, largely setting empirical research into the content of social structure to one side. Others doing empirical work have made very narrow theoretical claims that norms matter in this or that issue-area. There is no argument that norms in the various issue-areas might be patterned or related to one another in a coherent way.[46]

Jeffrey Checkel and Yale Ferguson have raised the same point: empirical applications of the constructivist approach have been seriously hampered by a lack of theory.[47]

The second set of objections that has been raised against the ways in which constructivists have proceeded until now is of a methodological nature. Paul Kowert and Jeffrey Legro have pointed out a number of methodological problems that have plagued empirical research built on constructivist assumptions.[48] First, they have argued that constructivists have not always been able to distinguish clearly between the norms of international actors and the effects of these norms. Furthermore, they have noted a bias in constructivist research towards the norms and values that quite obviously 'worked'. Norms or values that could have become influential, but in the end were not, have been neglected. Part of this bias has been a focus on a single norm, value or belief, or at most on a very small number of these predispositions. Fully fledged cultural approaches, which look at interpersonally shared *systems* of norms, values and beliefs, have been undertaken far less. Third, Kowert and Legro have signalled the problem of the 'ubiquity' of norms. People are often assumed to have so many identities, or so many different aspects to their identity, that in empirical research it always seems possible to construct some *ad hoc* explanation of behaviour in terms of a norm or value that the involved people have adhered to.

In my view, one reason for the existence of these three methodological problems is the inductive way in which scholars have tried to focus on norms and values. As objects of their empirical analyses constructivists have often taken concepts that are very familiar to the involved actors. For instance, research has focused on the 'chemical weapons taboo', or the norm of 'humanitarian intervention', or on the spread of 'Stalinist economic policies'. These are concepts that are often used, in speech and in writing and therefore probably in their thinking, by the involved people. As these concepts are so dear to international actors, it becomes difficult to distinguish between these norms and their effects. Also, it is quite difficult to relate such highly specific values and norms to other predispositions without the help of an existing body of theory. As a result, constructivist research has sometimes remained limited to a single norm or value, or at the most to a quite limited number of norms or values. Furthermore, international actors are often subject to many societal forces. They are confronted with many different beliefs, norms and values. This makes it possible for the inductive analyst to always find a norm, value or belief that can explain behaviour.

A possible solution to the methodological problems highlighted by Kowert and Legro is to delineate norms and values in a more deductive manner than in the inductive way that has been followed most often. This more deductive method would consist of distinguishing norms and values a priori, for instance on the basis of an existing theory about the mindsets of actors. Outside the study of IR, many empirical studies have been undertaken on norms that have been formulated in a more deductive way. In the field of comparative politics, one of the best-known examples is the work on 'civic culture' by Almond and Verba.[49] In this work, Almond and Verba explain the stability of democracy within a country in terms of the precise mix of three ideal-typical political cultures (parochial, subject and participant) that exists in the country. Another influential example is Geert Hofstede's cultural analysis.[50] In his research, Hofstede relates a wide variety of practices and opinions to four basic dimensions of national culture: power distance (the issue of human inequality), uncertainty avoidance, individualism and masculinity. In the field of ethnic studies an interesting example has been provided by Marc Ross, who has accounted for differences in the level of violence between societies on the basis of norms suggested by psychoanalytic theory (especially norms concerning child-raising).[51]

By following a more deductive approach, the conceptual problems stated above would emerge less easily. Norms and values distinguished on the basis of a priori theorizing are often quite different from the concepts and words used by the involved actors. This would make it easier to distinguish between norms and their effects (on other ideas or behaviour). It would also permit a focus on political or organizational cultures, i.e. on systems of related norms, values and beliefs that are widely shared within a polity or organization. Last, in this way, the norms and values that are expected to be important are announced before starting an empirical investigation. Thus, it should be easier to resist the temptation of using any norm no matter what to explain behaviour. Grid-group theory could provide an a priori formulated typology of widely held norms, values and beliefs that influence international processes. If applied in such a way, constructivist theorizing could be undertaken which does not suffer so much from the methodological problems raised by Kowert and Legro.

Grid-group theory may therefore be helpful in addressing some theoretical as well methodological problems that have plagued constructivism. Both sets of problems spring from the same source: absence of constructivist theory. Grid-group analysis not only meets the constructivist requirements, but also provides an interesting typology for the analysis of social processes that offers a basis for making and testing general statements.[52]

Conclusion: possible uses of grid-group analysis in (international) political theory

This chapter has argued for the importance of applying grid-group theory in studies of international politics. In the course of the discussion, four possible uses of grid-group analysis in (international) political theory were mentioned. First, cultural theory provides an elaborate taxonomy of the preferences, perceptions and values of groups of people that might be helpful in empirical research by making cases comparable as well as by instructing researchers as to which information they are looking for. Moreover, the theory has a predictive/explanatory element: with the help of the analysis, people's policy preferences and behaviour within an issue-area could be explained in terms of their underlying cultural bias. Third, cultural theory might help in explaining international change. By setting out ways in which social change may occur, grid-group theory also offers a partial answer to the agent-structure problem. Last, grid-group theory might also be helpful in strengthening the '(self-)reflexivity' of political theorists. Of course this does not mean that cultural theory is without its own limitations and flaws. Important conceptual challenges remain. The approach is still very much in development.

Grid-group theory might be usefully employed in various subfields of IR. This chapter will finish by suggesting some of the ways in which cultural theory could be applied to a few such subfields of IR. When used in foreign policy analysis, grid-group theory could provide a typology of four different belief systems that influence politicians' foreign policy decisions. Each of these belief systems would need to be based on the preferences for organizing social relations, the views of (human) nature, the ways of assigning blame when things go wrong and the preferred methods of matching needs and resources that grid-group analysis spells out for each way of life. On the basis of these predispositions, it should be possible to delineate four distinct belief systems that politicians might adhere to, each consisting of a different view of how international relations should be structured, an alternative belief in the feasibility of international cooperation and a definition of, and solution to, specific foreign policy issues. It would be interesting to test whether these belief systems provided more powerful explanations of important foreign policy decisions than the belief systems suggested by other writers, for instance the operational codes developed by Holsti and Walker.[53] In the study of the organizations involved in the international system, cultural theory might help to elucidate some aspects of the links between the ways in which these organizations are structured and the views of the officials working within them. In conflict and peace research, grid-group theory might help to clarify the origins of disagreements, misunderstandings and communication breakdowns between people. It might also contribute to a further understanding of political violence, since such violence seems to be related to the strength of group solidarity and the willingness of people to sacrifice themselves for others, as well as actors' perceptions of human nature and 'the Other'.[54] In international political economy, grid-group theory might contribute to an understanding of the alternative beliefs that exist with regard to the ways in which trade and monetary relations should be structured, as well as the best paths to development. In this respect, it is interesting to note that the four ultimate values which, according to Susan Strange, direct the choices of actors in the international political economy (wealth, freedom, security and equality) can easily be associated with the three active ways of life pointed out by grid-group theory. Wealth and freedom are major concerns of individualists, security and order are especially important to hierarchists, and equality is the main political goal of egalitarians.[55] Grid-group analysis may also make important contributions to international regime theory. These contributions are described in the next chapter.

3
Regimes, Institutions and Four Cultures

In Chapter 2 the argument was made that grid-group theory should be extended to the study of international relations. The present chapter will attempt to do just that and will show how three hypotheses concerning transboundary relations can be derived from a combination of international regime theory, new institutionalism and cultural theory. First, a brief overview of the development of regime theory will be offered. In this section, regimes will be defined as international institutions. Thereafter, the concept of 'institutions' will be clarified, and how regime analysis, several institutionalist insights and grid-group theory can be combined will be discussed. This will lead on to the general theoretical model of this study in the subsequent section. Next, parts of this general model will be specified by showing how four 'international ways of life' can be derived from cultural theory. Each of these international ways of life will include a preferred way of organizing international governance, conception of peace, trust in foreigners and a management system for international issues. In the final part of the chapter, three propositions will be introduced concerning international relations that follow from the model. The first two of these propositions are specifications of the basic claim that 'cultures matter in transboundary relations' and will be further explored in the case study of the environmental protection of the Rhine that is presented in Chapters 4 and 5. The third proposition highlights one way in which 'institutions matter in transboundary relations', and will form the basis for the empirical analysis undertaken in Chapter 6, in which efforts to protect the environment of the Rhine are compared with attempts to preserve the Great Lakes ecosystem.

International regimes and regime theory

Initially, regime theory[1] was developed partly in an attempt to capture the cognitive elements that influence international cooperation. None of the approaches represented in the inter-paradigm debate had adequately included the role of perceptions, ideas and values in their explanations.[2] By contrast, in the late 1970s and early 1980s scholars such as John Gerard Ruggie, Ernst Haas and Oran Young started to focus on how actors' perceptions and values interacted with material circumstances to produce international institutions.[3] This emphasis on the role of human understanding in the formation of international regimes was captured in the authoritative definition that Stephen Krasner presented of regimes as 'sets of implicit or explicit principles, norms, rules, and decision-making procedures around which actors' expectations converge in a given area of international relations'.[4] However, as Friedrich Kratochwil and John Gerard Ruggie have pointed out, in actual case studies of international regimes the focus has most often been on how 'material' or 'structural' factors (such as power capabilities of actors, monitoring activities, and the economic benefits and costs of cooperation) have influenced international regimes. In contrast, cognitive factors such as intersubjective understanding, information processing and communication have been neglected. Kratochwil and Ruggie argue that, as a consequence, important insights may have been missed concerning regime robustness and change, as well as with regard to the relations between regimes, international organizations and transboundary processes.[5]

More recently, Peter Haas has made a valiant effort to reincorporate a focus on cognitive elements into regime analysis with his epistemic community approach.[6] He defines epistemic communities as 'networks of professionals with recognised expertise and competence in a particular domain and an authoritative claim to policy-relevant knowledge within that domain or issue-area'.[7] These professionals have a shared set of normative and principled beliefs, shared causal beliefs, shared notions of validity and a common policy enterprise. Peter Haas, and others, have highlighted the conditions under which the ideas and values of these transnational networks of professionals can have a significant influence on regime formation. They have argued that, in today's world of complex interdependence, politicians and government officials do not have all the expertise required for a full understanding of many important policy issues. A case in point is the problem of ozone depletion. In situations like this, politicians and diplomats have to rely on the expertise of others (e.g. scientists, representatives from NGOs or international advisory councils), which sometimes gives these professionals significant leverage over international negotiations. Their values and beliefs then become embodied in international regimes.

Although the epistemic community approach has resulted in some interesting case studies of regime formation and change, and has brought cognitive factors back into regime analysis, the approach has also suffered from a conceptual weakness. Specifically, it has lacked an understanding of the ways in which international actors who are not members of an epistemic community think and act. In Peter Haas's approach, it sometimes seems as if only members of epistemic communities have normative and causal beliefs and notions of what counts as valid knowledge. All other actors, be they politicians, diplomats or other professionals, seem to lack these characteristics. In this sense, the epistemic community approach is wanting. A more comprehensive version of this approach would not only spell out the values and beliefs of epistemic community members, but also highlight the values and opinions of the other actors involved in the formation and development of an international regime.

I believe that grid-group theory can be of great help in this respect.[8] When applied to the formation of international regimes, it could make predictions about the norms, goals and beliefs of *all* actors involved in the issue-area. It would portray regime formation and development as a struggle between actors who follow the hierarchical, individualistic, fatalistic or egalitarian ways of life. These actors would be represented in government delegations, firms, international organizations, citizens' groups, the media and so on. The content of an international regime would be decided by the leverage that these actors had over each other, and the extent to which the actors were willing and able to find acceptable creative solutions to the policy problems. Such an analysis would be compatible with Peter Haas's epistemic community approach, because it would similarly show that, in certain highly complex issue-areas, government officials and diplomats sometimes have to rely on the expertise of other actors, whose norms and beliefs then start to influence the content of the international regime. The major difference from the epistemic community approach (as it now stands) would be that an application of cultural theory to international regimes could specify the values of both government representatives and all other involved actors.

The remainder of this chapter will extend grid-group theory to the analysis of international regimes in this manner. On the basis of the theory, four alternative systems of thought and action will be constructed concerning the questions of how international issue-areas should be regulated, what constitutes peace and fairness in the international realm, whether and how peace and fairness can be achieved in international relations, as well as preferences for specific policy options in issue-areas. In other words, *four alternative sets of principles, norms, rules and procedures for international issue-areas* will be delineated. These four 'international ways of life' should be seen as *ideal types*. It will be assumed that both the thought and behaviour of actors within an international issue-area to a significant degree tend to overlap with one of these international ways of life. By welding together cultural theory and regime analysis in this way, it should be possible to make a contribution towards meeting the challenge formulated by Kratochwil and Ruggie, i.e. to pay more attention to the role of cognitive factors in the formation of international regimes.

If regime analysis were to be combined with cultural theory, the former approach would also be better poised to meet two other objections. The first of these objections has been laid down by James Keeley, who has argued that regime analysts have focused solely on the norms and beliefs that are embodied in international regimes and have neglected the norms and beliefs that are excluded from regimes.[9] Grid-group theory would specify the various norms and beliefs that all the different actors within an international issue-area strive to get accepted. This would offer a basis for establishing which norms and beliefs have become widely followed within an international issue-area with those that have not become dominant. A second objection has been raised by Susan Strange.[10] She has criticized regime analysts for employing ill-defined and vacuous terms, such as principles, norms, rules and procedures. Again I believe that cultural theory could be helpful here. By postulating precisely which alternative sets of principles, norms, rules and procedures different actors within an international issue-area tend to prefer, it could bring more conceptual clarity to regime analysis.

Krasner's well-known definition of international regimes (see above) very much resembles a more recent one, formulated by Marc Levy, Oran Young and Michael Zürn. They write that international regimes are 'social institutions consisting of agreed upon principles, norms, rules, procedures and programmes that govern the interactions of actors in specific issue-areas'.[11] The latter definition makes explicit that international regimes should be seen as institutions. In following this view in this research, the terms 'regimes' and 'international institutions' will refer to one and the same thing. Defining regimes as institutions opens up the possibility of linking regime theory with institutional frameworks developed in other branches of the social sciences. It also begs the question what exactly is meant by the concept of institutions. This is the topic of the next section of this chapter.

Institutions and the new institutionalism

Institutions are self-sustaining social patterns. In the words of Ronald Jepperson, their persistence is

> not dependent, notably, upon recurrent collective mobilization, mobilization repetitively reengineered and reactivated in order to secure the reproduction of a pattern. That is, institutions are not reproduced by 'action', in this strict sense of collective intervention in a social convention. Rather, routine reproductive procedures support and sustain the pattern, furthering its reproduction – unless collective action blocks, or environmental shock disrupts, the reproductive process.[12]

In other words, institutions are patterns of thought and behaviour that are taken for granted. Most people accept these patterns as the way in which things are done.[13]

Since the middle of the 1980s, renewed attention has been given to the functions and determinants of institutions within the social sciences in general. The various theoretical approaches with which the roles of institutions have been studied since this time have been loosely grouped under the term 'the new institutionalism'.[14] Regime analysis can be seen as the contribution that the study of international relations has made to this new institutionalism.[15]

Peter Hall and Rosemary Taylor have distinguished between three different schools of thought within new institutionalism: historical institutionalism, rational choice institutionalism and sociological institutionalism.[16] Below is a brief description of these schools of thought, so as to be able to clarify the use of the term 'institutions'.

Historical institutionalists define institutions as the formal or informal procedures, routines, norms and conventions embedded in the organizational structure of the polity or economy. They can range from the rules of a constitutional order or the standard operating procedures of a bureaucracy to the conventions governing trade union behaviour or bank–firm relations. Four features of this new institutionalist approach stand out. First, historical institutionalists tend to conceptualize the relationship between institutions and individual behaviour in relatively broad terms. Institutions form the background against which individual actors behave. Second, historical institutionalists emphasize the asymmetries of power associated with the operation and development of institutions. Rather than posit scenarios of freely-contracting individuals, they are more likely to assume a world in which institutions give some groups or interests disproportionate access to decision-making processes. Third, historical institutionalists tend to have a view of institutional development that emphasizes path-dependency and unintended consequences. In their eyes, institutional settings are shaped at critical historical junctures. These critical junctures often involve clashes of alternative discourses, cultures or ideologies. Once the dust of these clashes has settled, and a set of institutions has been formed, the institutions tend to be perpetuated. The strategies induced by a given institutional setting may ossify over time into worldviews, which are propagated by formal organizations and ultimately shape even the self-images and basic preferences of the actors involved in them. Fourth, historical institutionalists are especially concerned to integrate institutional analysis with the contribution that other factors, such as ideas, can make to political outcomes. They rarely insist that institutions are the only causal force in politics. They typically seek to locate institutions in a causal chain that accommodates a role for other factors, notably socioeconomic development and the diffusion of ideas.

Rational choice institutionalists, who make up the second school of new institutionalism, see politics as a series of collective choice problems. Actors have a fixed set of preferences or tastes, behave instrumentally so as to maximize the attainment of these preferences, and do so in a highly strategic manner that presumes extensive calculation. Institutions are viewed as voluntary agreements that allow actors to overcome collective choice problems. Often, institutions achieve this effect by lowering information costs. Within the study of international relations, this form of the new institutionalism has come to be known as neoliberalism or neoliberal institutionalism (discussed in some detail in Chapter 5).

Sociological institutionalists (the third school) tend to define institutions much more broadly than political scientists do to include not just formal rules, procedures or norms, but the symbol systems, cognitive scripts and moral templates that guide human action. Such a definition breaks down the conceptual divide between 'institutions' and 'culture' and the concepts shade into each other. According to sociological institutionalists, institutions influence behaviour by providing the cognitive scripts, categories and models that are indispensable for action, not least because without them the world and the behaviour of others cannot be interpreted. Institutions affect behaviour not simply by specifying what one should do, but also by specifying what one can imagine oneself doing in a given context. In this view, institutional diffusion is supposed to take place through imitation.

The way in which grid-group theory has usually been presented would make it a form of sociological institutionalism.[17] Usually, the four ways of life (each consisting of both a cultural bias and a corresponding set of social relations) have been portrayed as institutions in and of themselves. Here I somewhat diverge from this practice.[18] My understanding of institutions falls more within the historical version of the new institutionalism.[19] All four features of the historical institutionalism that have been highlighted by Hall and Taylor appear in the theoretical model that is used in this study. In my conceptualization, institutions are background variables, different from the ways of life or cultures that actors adhere to. Furthermore, institutions influence the distribution of power and influence among actors. In addition, political life is seen as path-dependent and riddled with unintended consequences. Last, institutions are not the only causal variables in politics. Cultures matter as well.

The concept of institutions is used here in two ways. First, international regimes are viewed as institutions. Actors within an international issue-area are expected to favour alternative sets of principles, norms, rules and procedures for dealing with the transboundary issue at hand. These actors strive to get their preferred way of life adopted by other people and organizations within the issue-area. At certain, 'critical' junctures in time, this cultural clash is settled in favour of some organizations at the expense of others. The resulting international regime consists of those principles, norms, rules and procedures that have been advocated by the 'winning' actors. Other organizations and individuals may not agree with, or morally condone, the principles, norms, rules and procedures that make up the international regime, but they will have to learn to live with them. In the course of time, these other actors are expected to do so. The principles, norms, rules and programmes of the most powerful or persuasive actors then become institutionalized, ingrained.

The second way in which the concept of institutions is employed here can be formulated thus. In this conceptualization, actors adhering to alternative ways of life attempt to get their preferred principles, norms, rules and procedures adopted in an international issue-area. This struggle will be influenced by domestic and international institutions. Such institutions will especially shape *the relations between actors adhering to different ways of life*. For instance, domestic and international institutions determine the access that actors have to the processes of information gathering, problem definition as well as policy choice. Some institutions give all actors access to decision-making, while other institutions privilege only a few actors. Different institutions may also skew the distribution of power resources between actors. Organizations and individuals may use all kinds of power resources to get their preferred principles, norms, rules and programmes adopted, including money, expertise, violence, rhetoric, etc. Domestic as well as international institutions affect the balance of power between actors following alternative ways of life. A last example concerning the way in which institutions shape the relations between actors espousing alternative ways of life has to do with mutual respect and understanding. Chapter 6 will show that a number of institutions tend to create more understanding between organizations and individuals following different cultures. Under these 'consensual' institutions, the ways of life to which actors adhere are still at odds, but the differences tend to be smaller. Mutual understanding and respect between actors following alternative cultures occur. Synthetical policy innovations, satisfying all perspectives, are apt to be found more easily under such institutions. Chapter 6 will also highlight a set of institutions that tends to divide and polarize actors believing in alternative ways of life. Under these more 'adversarial' institutions, the alternative perspectives that actors bring to bear on political issues tend to lie further apart. Cultural struggle will be more intense under such conditions.

By shaping the links between actors espousing alternative cultures in various ways, institutions co-determine international outcomes. The domestic institutions that have this function include the constitution of a country, the way in which the political economy is organized, the political role of judges, the standing that citizens have in courts of law, and so on. Chapter 6 will look at a variety of institutions influencing the links between the executive, legislature, judiciary, firms and citizens' groups. International institutions (alias regimes) also feature in the analysis. Regimes are not only the outcomes of cultural struggles. Once established, they also tend to impact on the efforts of actors to impose their preferred way of life within a transboundary issue-area. For instance, a regime principle may be to give NGOs access to intergovernmental negotiations. This principle will strengthen the position of NGOs within an international issue-area and will increase the likelihood that their viewpoints will be taken up by the governments involved.

The relationship between (domestic and transboundary) institutions on the one hand and cultures on the other is therefore not one-way. Institutions influence cultural battles in two ways. Institutions shape the links between actors adhering to various ways of life. This can be called the *regulative* impact of institutions. In addition, once installed, institutions will become increasingly incorporated into the way in which actors define themselves and perceive the world. This can be dubbed the *formative* impact of institutions. But both domestic and international institutions themselves are also the outcomes of cultural battles. Alternative cultures include different visions of the proper relations between the executive, legislature, judiciary, firms and citizens' groups. Actors adhering to different ways of life strive to get their viewpoints adopted by others. At particular, 'historic' points in time, these battles are decided in favour of one cultural perspective or a combination of perspectives. In sum, cultures and institutions mutually influence each other. Which set of elements should be seen as the independent variables and which set as the dependent variables simply depends on the research question. Chapters 4 and 5 will show how the international regime for the protection of the Rhine has been forged through cultural battles. In these chapters 'ways of life' are the independent variables and the principles, norms, rules and procedures of the Rhine regime the dependent variables. Chapter 6 will argue that (especially the domestic) institutions regulating the relations between actors following different ways of life (firms, governmental agencies, environmental groups) in the Great Lakes basin have been quite different from the institutions shaping the links between organizations espousing different rationalities within the Rhine watershed. It will also be shown that these institutional differences can account for why the toxicity of industrial effluents into the Rhine has been much less than the toxicity of industrial effluents into the Great Lakes basin. In Chapter 6, therefore, institutions are the independent variables, cultures are the intermediate variables, and the toxicity of industrial effluents (which can be seen as part of an international regime) the dependent variable.

By definition, international institutions are self-sustaining. This begs the question: how then can international regime change be accounted for?[20] Several possibilities to explain regime change exist. First, institutions within a certain issue-area may be affected by changes in other (international and domestic) issue-areas. A very simple example is a war breaking out between countries, disrupting environmental regimes in which the governments and firms of these countries have participated. In general terms, changes outside an issue-area may change the power resources, perceptions and preferences of the participants in the issue-area. Furthermore, regime transformation may be realized through policy learning and 'unlearning', accommodation and strife between actors, and technological developments within the issue-area itself. Again, it can be hypothesized that different regimes allow for alternative dynamics. Finally, there is the regime change that is stipulated by grid-group theory (described in Chapter 2). If a number of organizations which adhere to a certain cultural bias has been dominant within an international issue-area for an extended period of time, it sometimes happens that social problems start to emerge that can neither be understood nor solved with the help of that cultural bias. When these problems accumulate, people and organizations slowly start to switch to other cultural views. Regime adaptation will be the end result of this.

The theoretical model of this research

This section draws together the relationships sketched above. Taken together, these relationships constitute the theoretical model of this research in general terms, and can be illustrated with the help of Figure 3.1.

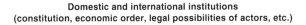

Domestic and international institutions
(constitution, economic order, legal possibilities of actors, etc.)

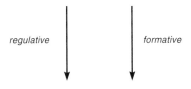

regulative   *formative*

Actors espousing alternative ways of life

cultural battle

Regime principles, norms, rules and procedures

Figure 3.1   The theoretical model

The four ways of life spelled out by grid-group theory are assumed to inform the principles, norms, rules, procedures and programmes that people within such organizations as governmental departments, research institutes, human rights organizations, environmental pressure groups, international organizations, agricultural and industrial firms, local government levels, etc. would like to realize within an international issue-area. Through their interaction, these organizations decide on the content of the principles, norms, rules, procedures and programmes that constitute a regime. Domestic as well as international institutions themselves influence this whole process in two ways. They structure the interaction between the involved organizations (the regulative aspect of institutions), while at the same time sustaining dominant ways of life (the formative aspect of institutions). Fundamental change within an international regime is expected to be primarily caused by the accumulation of social problems that cannot be dealt with on the basis of the dominant culture(s).

How does grid-group theory figure exactly in the theoretical framework of this book? The theoretical model is concerned with the actors that are involved in an international issue-area. (The empirical focus is on the environmental protection of transboundary watersheds, but the model could be applied to any international issue-area.) Grid-group theory spells out the principles, norms, rules and programmes that different groups of actors try to follow for themselves and strive to get accepted by other actors involved in the issue-area. This is the main way in which cultural theory is applied. These competing principles, norms, rules and procedures favoured by different groups of actors within an international issue-area are called their 'cultures' or 'international ways of life'. However, this is not all. The power struggle between actors adhering to different cultures are channelled and structured by a host of domestic and international institutions. These are the institutions that shape the relations between actors by distributing power resources and making mutual respect (im)possible (among other things). These domestic and international institutions influencing the relations between the organizations within an international issue-area are themselves the outcomes of previous (and ongoing) clashes between actors with different worldviews. In principle, therefore, grid-group theory could also be used to elucidate the development of these institutions. This is the case as cultural theory can be applied at any level of analysis. The conclusion to Chapter 7, for instance, briefly hints at the cultural origins of the domestic institutions that have shaped the relations between the different cultural actors involved in the protection of the Rhine and the Great Lakes. This is another way in which grid-group theory could be employed. Nevertheless, if we wish to pursue research on a single international issue-area, it may often not be feasible, or useful, to delve deeply into the origins of all the domestic and international institutions that shape the links between the persons and organizations active in the issue-area. This is so as many of these institutions (such as the constitutions of the countries concerned) have often been formed in cultural clashes that have taken place outside of the international issue-area, and in different time periods than the one we are interested in. In that case, a researcher on international regimes is best advised to use grid-group theory mainly to find out which 'cultures' or 'international ways of life' the organizations involved in the issue-area believe in: their preferred principles, norms, rules and programmes with regard to the transnational issue at hand. The institutions shaping the links between these different cultural actors can then often be better treated as background variables, and be taken from the existing historical-institutional literature. This is the strategy followed in Chapter 6.

It is not in any way assumed that international regimes are always harmonious – islands of stability in a sea of international struggle. There has been a tendency in the literature to make this assumption. First, that there will always be pockets of resistance to the principles, norms, rules and procedures that prevail within an international issue-area is to be expected. There are always actors who radically disagree with the dominant discourse within an international issue-area. These actors are the defenders of alternative faiths, and thereby the carriers of social change. Second, the principles, norms, rules and programmes that are prevalent within an issue-area may be conflictual in themselves. For example, 'distrust foreigners' or 'arm yourself to the teeth' may be dominant principles within a transboundary issue-area. Widespread adoption of transboundary principles, norms, rules and procedures does not necessarily mean international harmony.[21]

What still remains to be done, before the empirical research of this study can be presented, is to spell out the exact principles, norms, rules, procedures and programmes with regard to international issue-areas that are part of the four ways of life of grid-group theory, as well as to present three propositions that can be derived from the above theoretical model. This will be undertaken in the last two sections of this chapter.

Four international ways of life

The four ways of life distinguished in cultural theory can be extended to include principles, norms, rules and decision-making procedures that constitute international regimes. Below will be shown that the adherents of individualism, hierarchy, fatalism and egalitarianism favour quite divergent international principles, norms, rules and procedures. In this exercise, principles and norms, as well as rules and decision-making procedures, are grouped together, first because it is quite difficult to disentangle principles from norms and rules from procedures, and second because it is rather straightforward to separate principles and norms on the one hand from rules and procedures on the other. The former group of dispositions is more abstract and general than the latter set.

The four alternative sets of international principles, norms, rules and procedures that will be presented should be seen as *ideal types*. Their specific content can be expected to vary somewhat from time to time and place to place. Also, in real life the behaviour and mindsets of people will probably to some degree consist of a mix of several ways of life, though for the theory to be useful people should in the main be disposed to a single way of life (at least with regard to a separate issue-area). This makes a clarification concerning terminology necessary. When 'individualists', 'egalitarians', 'hierarchists', or 'fatalists' are referred to it will mean actors who fully adhere to the ideal types that will be presented. Yet, in reality many actors (be they organizations or individuals) combine various ways of life to some degree.

The elements of the international ways of life have been chosen somewhat arbitrarily. Principles, norms, rules and procedures have been selected that are regarded as important in the existing IR literature. But many other elements could also be included. The four international ways of life that are presented below should therefore not be seen as closed entities.

In conclusion, the argument is that regime formation and development can fruitfully be seen as a struggle between actors who by and large follow the hierarchical, individualistic, fatalistic or egalitarian ways of life. These actors can be (officials of) government delegations, NGOs, international organizations, business firms, the media and so on. The content of an international regime will be decided by the leverage that these actors have over each other, and the extent to which the actors are able to find either compromises or synthetic solutions to policy problems. In other words, the actual principles, norms, rules and procedures that make up an international regime will be the outcome of a clash between organizations striving to have their own preferred set of principles, norms, rules and procedures accepted and implemented. The content of these competing sets is described below.

Principles and norms

Within regime theory, principles are usually defined as 'beliefs of fact, causation, and rectitude'. Norms are 'standards of behaviour defined in terms of rights and obligations'.[22] At least four principles and norms can be spelled out for each way of life.

The essence of international issues

In collective choice analyses, the provision of solutions to social issues and needs are dubbed 'goods'. Often, two characteristics of such goods are deemed essential: *jointness of consumption* and *exclusiveness of consumption*. Jointness of consumption denotes the degree to which use of a good by a person still leaves the availability of that good to others. Exclusiveness of consumption stands for the extent to which it is possible to exclude individuals from benefiting from the provision of a good. By assigning two values (high and low) to either of these properties, four types of social goods can be distinguished: collective, or public, goods (high jointness, low exclusiveness), private goods (low jointness, high exclusiveness), common-pool resources (low jointness, low exclusiveness) and club goods (high jointness, high exclusiveness).[23] Usually, the characterization of a domestic or international issue as a collective good is treated as a purely technical exercise, based on objective criteria and fixed properties of the issue at hand. However, it has also been argued that jointness and exclusiveness should not be regarded as natural, immutable properties of goods, but instead should be seen as social constructs.[24] Albert Hirschman put it thus:

> He who says public goods, says public evils. . . . What is a public good for some – say a plentiful supply of police dogs and atomic bombs – may well be judged a public evil by others in the same community.[25]

In other words, whether a societal issue can be described in terms of one of the above mentioned 'goods' is dependent on the perceptions and actions of the actors involved in the issue. With the help of grid-group analysis, it is possible to postulate the affinities that different cultures have with alternative ways of perceiving the essence of international 'goods' or issues.

Hierarchists will tend to perceive many international issues as collective goods. Due to the non-excludability and jointness of collective goods, no one will want to contribute to their production and the market will therefore 'underprovide' such goods. Only governmental provision will be able to remedy this situation. Therefore, the existence of collective goods strengthens the call for governmental action and thus supports the hierarchical way of life. In the international realm, such governmental action will usually have to consist of extensive intergovernmental cooperation.

Individualists will tend to characterize international issues as private goods. No extensive intergovernmental action is needed for the efficient provision of such goods. Open international markets and mutual trust between parties to a contract are all that is needed. Individualists believe that organizations and firms will often be able to find ways in which to reduce jointness of consumption and increase excludability, thus turning solutions to international issues into private goods.

Egalitarians will warn us that many worldwide concerns should be seen as common-pool resources. The strict limits of common-pool resources, in combination with their non-excludability, will make depletion of such resources imminent. The only solution is voluntary constraint on the part of all domestic and foreign organizations and individuals that are involved. This reduced consumption will diminish the differences in wellbeing and status between people, which is the ultimate aim of egalitarians.

Fatalists feel excluded from society and regulated from without. In their view, many international issues are club goods. Such goods are characterized by both a high degree of excludability and a high degree of jointness of consumption. This means that a select (though not necessarily small) group of organizations and people will be able to benefit greatly from the provision of such goods, while others will be left out. Fatalists feel that they are the ones who are often left out.

Perhaps a brief example will illuminate the above. Consider the classical example of the construction of a lighthouse in a coastal region with a large port. Hierarchists will tend to argue that the construction of such a lighthouse constitutes a collective good: everyone who uses the coast or who lives in the coastal region will be able to benefit from it, while no one can be excluded from the benefits that it confers. Therefore, they will argue, people, by themselves, will not be willing to contribute to the financing of the lighthouse and government has to step in.

Individualists will see ways of turning the construction of the lighthouse into a private good. For instance, they can argue that ships of countries or firms that have not contributed to the costs of the lighthouse will be refused right of passage through the territorial waters. They may also redefine the problem of the building of a lighthouse altogether. Individualists may argue that the underlying need is not so much for a lighthouse, but for ways of navigating ships safely through the night. The construction of a lighthouse may be one solution to this, but others can also be conceived. For instance, the development of new radar equipment and the use of pilots may also solve the problem. Markets exist for radar equipment and pilots, so no government regulation will be needed.

Egalitarians will argue for a communal decision on the construction of the lighthouse. In addition, they will warn of the many common-pool resources that will be depleted as an indirect effect of the building of the lighthouse. The lighthouse will spoil the natural outlook of the beach and will consume lots of energy. Oil spills and other problems related to increased shipping may destroy the fragile ecosystems of the coastal region.

Fatalists living in the region will not really want to contribute to the financing of the lighthouse. They believe that they will not receive any of the economic benefits that will be reaped from the project. These will all go, fatalists argue, to the owners of the construction company, shipping magnates and bankers.

Ideals of peace[26]

Egalitarians will tend to have a broad view of international peace. Their definition will include not just the absence of violence, but also the existence of equitable relations between peoples, as well as between humankind and its natural environment. In this view, peace encompasses the absence of wars, environmental problems and large economic differences between nations. Egalitarians will argue that such a peace will be secured if humankind reorganizes itself in small-knit, autarchic and free communities. The relationships between these communities would be limited, but friendly and voluntary.

Individualists will tend to see peace as the absence of wars as well as any other restrictions on personal freedom (including trade barriers). Such a peace could be secured by creating an open international economy which would foster specialization between countries. The resulting interdependence would constitute an important impediment to the outbreak of violence and economic strife between states.

The hierarchical view of peace will tend to include the absence of war, as well as widespread adherence by states to the traditional rules of international society.[27] The hierarchical peace ideal would be the establishment of a world government. Alternatively, they might opt for inducing a powerful state to act as a benevolent world hegemon that would secure peace in the international system. However, since the realization of either of these options often seems remote, hierarchists will tend to settle for a world system that is extensively regulated by intergovernmental agreements and international organizations.

Fatalists do not believe that a more peaceful international system can be achieved. Therefore, it does not make sense to them to form an opinion on how a peaceful world might look or how it could be achieved.

Governance of the international system and regimes

World government characterizes the ideal world order of hierarchists. Central planning and allocation is their favoured way of organizing both domestic and transnational social relations. However, the establishment of strong supranational organizations seems rather unlikely at present. For this reason, and also because hierarchists are pessimistic about the possibility of international cooperation (see below), they must settle for a second-best world order. This second-best choice is regulation and control of the international system through extensive intergovernmental cooperation and consultation. Governments should at a minimum adhere to the time-honoured principles of international society (e.g. formal equality of states, non-interference), as well as to the established practices and rules of international public law. Ideally, international regimes should be based on explicit agreements between state authorities which impose these regulations on their citizens.

In contrast, individualists would like to keep governmental regulation to a bare minimum, both at the domestic and at the international level.[28] They believe that unfettered competition in open world markets is a much more efficient allocation mechanism than regulation by governments and international organizations. Traces of the individualistic view of how international relations should be governed can be found in the vision propounded by chief executives of multinationals:

> Their vision . . . involves a world in which business corporations have replaced the nation-state as the effective unit of economic policy and resource allocations. There is nothing particularly international or global about the chairman of Dow Chemical's dream of establishing the world headquarters of the Dow Company on the truly neutral ground of . . . an island (owned by no nation), beholden to no nation or society.[29]

World government is anathema to individualists; international regimes should not be over-regulated by central authorities. Anarchy, in the sense of absence of world government, is not really a problem to individualists. They are actually more afraid of its opposite, extensive regulation by central authorities. The absence of world government is much more a concern for hierarchists.[30]

Egalitarians often distrust state authorities, and therefore also intergovernmental agreement. National and international governance should be formed in an open, free dialogue between all citizens concerned with the common good. Ideally, no special status should be given to representatives from central authorities or business organizations. International regimes should be voluntarily agreed on by all affected and concerned citizens, local and central authorities, and representatives of involved organizations from all countries.[31] According to egalitarians, the functioning of international regimes should not be hindered by existing national boundaries.

Fatalists do not believe that world order can come about through rational planning. They blame the anarchical nature of the international system for this. In anarchy it is impossible and dangerous to distinguish friends from foes. Long-lasting and extensive international cooperation is therefore unfeasible. Each state is unto itself, trying to cope with the vagaries of the international realm as best it can.[32]

Perceived feasibility of international cooperation[33]

The previous section dealt with how cooperation in the international system and international regimes should ideally be organized, according to the various ways of life. The present section is concerned with the belief held by the adherents of the four ways of life in the *feasibility* of establishing international cooperation.

Egalitarians will tend to believe that intergovernmental cooperation is not feasible. According to them, humans are born good but can easily be corrupted by large-scale institutions. They believe that relations between governments will tend to be coercive and conflict-prone. However, since egalitarians believe that humans are essentially good as long as they are not heavily regulated, they will also tend to believe in the feasibility of agreement among *citizens* belonging to different nationalities. Egalitarians will tend to view foreigners as their 'brothers' and 'sisters', with whom they want to live in peace and harmony.

Fatalists view human nature as essentially unpredictable. To be on the safe side, they feel inclined to distrust people and organizations. This view of human nature will induce fatalists to discount the feasibility of any form of international agreement. They will tend to think that such cooperation will be undermined by free-rider problems. In their view, governments cannot trust each other, and therefore cannot hope to establish international cooperation. As a consequence, nation-states seem to be entrapped in a system of conflict, rivalry and self-help.

The degree of confidence that hierarchists have in the possibility of international agreement is influenced by two factors. On the one hand, hierarchists assume that humans are essentially bad, but can be reigned in and redeemed by institutions. Therefore, they will tend to assume that cooperation between governments is not unfeasible. However, on the other hand, hierarchists (by definition) will be part of a strong and clearly separated and demarcated group of people with their own identity. Because of this, it seems reasonable to assume that hierarchists will tend to view international relations as a struggle between 'us' and 'them'. This perception will tend to decrease their willingness to cooperate internationally, as well as their confidence in the possibility of international accords.[34] Overall, therefore, hierarchists will tend to be rather pessimistic with regard to the feasibility of international cooperation, and rather distrustful of foreign 'opponents'.

The measure of trust that individualists have in international co-ordination is also subject to contradictory influences. Individualists believe that humans are self-centred and cannot be redeemed in any way. As a consequence, they will tend to assume that actors in international regimes will behave in a self-satisfying manner, thus making international cooperation more difficult to achieve. However, this view is mitigated by the inclination of individualists not to think in group terms, in terms of 'them' against 'us'. Moreover, in grid-group theory it is assumed[35] that individualists are rational enough to understand that their long-term interests are best served by not reneging on contracts and promises, by reciprocity and respect for the freedom of others. Last, individualists are typically *risk-takers*, not just on the stock market, but also with regard to business partners. They tend to assume that in the main actors will honour their promises and contracts. They expect that people will seek their own advantage in making deals. But individualists also tend to believe that once deals have been made, people will stick to them. If individualists lacked this kind of trust, then they could no longer justify their preference for social relations that are void of restrictions on personal freedom. Overall, therefore, individualists are relatively trusting of foreign actors and quite optimistic about the possibility of international cooperation.

Rules and procedures

Rules in international regimes are 'specific prescriptions or proscriptions for action'. Decision-making procedures stand for 'prevailing practices for making and implementing collective choice'.[36] On the basis of cultural theory, different preferences for rules and procedures within specific international issue-areas can be distinguished. Below, this will be illustrated for the (hypothetical) case of the environmental protection of a transboundary water basin.[37] Two sets of rules and procedures will be highlighted. The first set of rules and programmes has to do with how, when and which environmental goals should be reached. The second set concerns the question of how governments should induce firms to mend their polluting ways.

Establishing environmental goals

Individualists will at first be little concerned with the environmental protection of the international watershed.[38] To them, the basin is primarily an asset to be used for the purposes of consumption, production and trade. They will tend to ignore, or argue against, accumulating evidence of the ecological degradation of the watershed. Alternatively, they may believe that the sources of the environmental degradation could be swiftly dealt with. On these grounds, it seems only common sense to individualists that protection measures will only be taken after scientific research has conclusively shown that certain consumption and production practices are indeed harmful to the ecosystems of the basin. They will tend not to believe in the wisdom of the precautionary principle. Individualists will also insist on taking environmental measures only on the basis of cost–benefit analyses. In their eyes, the benefits of environmental measures (in terms of increased protection of the watershed) should always outstrip the costs (in terms of a decline in output and employment). Once individualists have accepted the need for environmental protection of the water basin, they will insist that this is done in a cost-effective and Pareto-efficient way. In their eyes, it does not make sense to make large investments in one place that only slightly improve ecological quality when it is possible to achieve a bigger ecological improvement for less money in another part of the watershed, even if this place is across a border.

Egalitarians will be the first ones to perceive environmental degradation of the transboundary basin. They will claim that the environmental degradation has already become widespread and is causing great harm to animals, plants and trees. In many present-day agricultural and industrial production processes, substances found in nature are chemically transformed. Lots of these substances will be discharged into the water. Egalitarians will tend to perceive all of these substances as potentially toxic. Therefore, they will argue for a strict application of the precautionary principle. A chemical substance should only be discharged into the water after it has conclusively been shown that this will not be harmful to the environment. As the development of new chemical substances far outpaces the progress of scientific measuring techniques, such a strict application of the precautionary principle would seriously cripple modern industry and agriculture. This will not be an overriding concern of egalitarians, who will try to broaden the issue of the environmental protection of the watershed into a general debate and critique of existing consumption and production processes. Only a fundamental change of existing agricultural and industrial practices, as well as of the profit motive, will ensure the environmental restoration of the watercourse. 'Ecocentrism' instead of 'egocentrism' is called for.

Hierarchists are interested in controlling and mastering nature on the basis of expert knowledge. In the past, this meant harnessing the water basin for productive uses by building dikes, dams, weirs, hydroelectric power stations, connecting canals and so on. These hierarchical engineering projects were heavily criticized by egalitarians who reject human subjugation of water basins.[39] In more recent times, hierarchists have turned their attention to environmental management. They tend to prefer elaborate, computer-guided systems with which to measure and regulate biodiversity, water quality and the water quantity of basins.[40] Again this does not quite square with the wishes of egalitarians, who generally prefer a 'hands-off' approach to nature. In principle, hierarchists will look sympathetically towards the precautionary principle. It is, after all, a hierarchical ambition to minimize risk on the basis of expert knowledge. However, they will not endorse a very strict application of this principle if this entails the immediate decline of modern agriculture and industry. Hierarchists want to minimize risks in order to maximize stability and order. A very strict application of the precautionary principle (such as that advocated by egalitarian groups) would lead to instability by disrupting industry. This will not do for hierarchists. Instead, they will put their faith in 'objective' risk analysis. In their view, experts are able to establish the objective risks of the presence of chemicals in natural systems, and are able to make balanced, cool-headed judgements about which environmental and health risks to take and which not. In the case of the protection of a water basin, this procedure will often lead to the selection of a select group of highly toxic chemicals which present large health hazards and which should therefore be banned.

It may be useful to compare the hierarchical approach to environmental risk with the approaches advocated by individualists and egalitarians.[41] Individualists tend to regard no chemical substance as harmful, unless the opposite has been *scientifically proven*. This is a risk-taking, or even risk-negating, approach. It leads to a very small list of chemical substances that are regarded as dangerous. Egalitarians tend to view all chemical substances as potentially toxic, unless *scientific proof* to the opposite is produced. This is a highly risk-averse attitude. It brings forth a very large list of chemical substances that should be avoided and banned. Hierarchists tend to accept that it is very difficult, if not impossible, to establish the exact toxicity of a chemical substance. Instead, they believe that it is possible for experts to calculate the *objective risks* to the environment and human health of releasing chemicals. These objective risk estimates pertain to two things. First, they rank different chemicals on a scale from harmless to highly dangerous. Second, these estimates pertain to the permissible concentration levels of each specific substance. Hierarchists therefore believe that they can gather objective, hard knowledge about which concentration levels of which specific chemicals pose the greatest risks to the environment. They will seek to keep the emission of these substances within the established limits. This is a moderately risk-averse approach. It usually leads to a list of suspected chemicals that is bigger than that of individualists and smaller than that of egalitarians.

Fatalists will not be worried too much about the environmental degradation. They will tend to feel that they are not able to either understand or influence both the ongoing degradation and the efforts to clean up the watercourse. They will be hoping that their practices will not be affected much by attempts to restore the environment of the watershed.

Business–government relations

Apart from their differences of opinion about when, how and which environmental goals to establish, the adherents to the different ways of life also disagree about how governments should influence firms to clean up their act.

Hierarchists will tend to believe in the usefulness of strict command-and-control policies for the protection of the watercourse. Industrial effluents into the waterway should be made subject to legal standards that are decided upon by the government. Government agencies should also check whether firms remain within these lawful limits. Part of the command-and-control policies espoused by hierarchists will be the central prescription of 'best available technical means' with which to treat wastewater.

Individualists will reject so much interference in business affairs by government. They will try to keep governmental regulation and prescription to a minimum. Individualists will put their faith more in 'gentlemen's agreements' or 'covenants' between business leaders and government officials. In such voluntary programmes, corporate executives pledge themselves to achieve certain environmental goals, perhaps in exchange for governmental support in other areas. In the hypothetical case at hand, corporate leaders can voluntarily agree to reduce the release of pollutants into the watershed. Individualists will tend to believe in bargaining between the public and private sectors. They will usually oppose governmental prescription of 'best available technical means' for the purification of effluents on the grounds that this is too static a concept that will stifle technological innovation. Other policy tools with which individualists will feel comfortable are market instruments. The release of toxic effluents into the waterway could for instance be taxed by governments. Also, permissions to discharge toxic substances into the basin could be made tradable. Such market-based instruments will tend to be favoured by individualists.

Egalitarians will face one of their perennial dilemmas. As a low grid culture, they will be tempted to reject extensive governmental regulation. Yet, at the same time, they want to bring about many social changes, including more respect – even rights – for animals and plants. Plus, they tend to distrust the intentions of huge business conglomerates. These two latter considerations pull egalitarians in the opposite direction: extensive government regulation of the polluters within the water basin. Because of their view of nature as extremely fragile, egalitarians may even end up preferring stricter command-and-control policies than are favoured by hierarchists. However, egalitarians will tend to see these policies as temporary solutions at best. Only a radical overhaul of society will ensure a sustainable use of the water basin.

Fatalists will not elaborate extensive plans for managing the eco-systems of the watershed.

An overview

The four international ways of life are summarized in Table 3.1. Items 1–4 in the rows are principles and norms that adherents of the various cultures like to see institutionalized in international regimes; the last two elements in the rows concern rules and programmes that the proponents of the ways of life like to see implemented in an international regime to protect an ecosystem such as a watershed.

Again it should be stressed that these four international ways of life should be treated as ideal types useful for describing the moti-vations and actions of organizations and individuals. Seldom will the belief systems of international actors fully overlap with a single way of life. However, it is assumed that the views of actors often overlap to a significant degree with one of these cultures. In addi-tion it needs to be realized that these four ways of life can be expected to vary somewhat from time to time and place to place.

These four perspectives are assumed to direct the actions and opinions of those who are active within an international issue-area. Regime formation and development is therefore presented as the struggle and interaction between actors adhering to different inter-national ways of life. Domestic and international institutions channel these conflicts by allocating (legal, political, financial) resources among the actors espousing alternative rationalities. The principles, norms, rules and procedures that make up an international regime are the outcomes of this struggle and interaction.

In the empirical part of this book the actors described in this vein will usually be groups of organizations (such as firms, farms, government agencies and environmental groups). When one such group of organizations is called for instance 'egalitarian', it will merely mean that this collection of organizations comes closer to the egalitarian ideal type than other actors.

Who's afraid of international anarchy?

Each of the above international ways of life contains a different way of reacting to the absence of world government. International anarchy affects the behaviour of hierarchists much more than the behaviour of individualists and egalitarians or even fatalists. Egalitarians belong to a low grid culture, and tend to distrust cen-tral authority. They can be concerned with the absence of world

Table 3.1 Four international ways of life

	Hierarchy	Individualism	Egalitarianism	Fatalism
1. *Perception of international issues*	Public goods	Private goods	Common-pool resources	Club goods
2. *View of international peace*	Absence of war, plus non-intervention and other principles of public international law	Absence of war Openness of borders for trade	Absence of war, plus equitable political, economic and ecological relations around the world	Absence of war
3. *Preferred system of international governance*	Official inter-governmental agreements and treaties	Allocation through transnational markets	Extensive citizen participation in decision-making	Self-help
4. *Belief in possibility of international cooperation*	Low	High	Low regarding intergovernmental cooperation High concerning cooperation between citizens	Very low
5. *How to decide on environmental goals*	Precautionary principle Objective risk analysis	Against precautionary principle For scientific certainty For cost–benefit analysis	Very strict precautionary principle	No plans
6. *How to achieve environmental goals*	Best available technical means Command-and-control policies	Voluntary agreements Market instruments No best available means	Best available technical means Command-and-control policies Fundamental social change as lasting solution	Acceptance of perils

governance, as they often see the present anarchical international system as a place where state monolith meets state monolith to play power games or to broker deals – without much regard for the common people. However, they have the same sceptical attitude towards domestic policies of state authorities, and therefore behave similarly in domestic and international realms. Individualists want to build up personal networks through which they can truck, barter and exchange. It does not matter to them whether these networks extend across borders or not. As they reject central planning at the domestic level, they are certainly not willing to endorse central authority at the international level. In sum, individualists too behave no differently across or within borders. The same goes for fatalists. They see no opportunities for cooperation and planning in the international realm, but they also hold similar views within domestic societies, where they usually occupy marginalized positions.

In contrast to the other cultures, the behaviour of hierarchists is quite different at the international level than nationally. When faced with a domestic social problem, a hierarchist will propose a whole battery of sophisticated, expert-based methods and programmes with which to study and measure the problem, as well as solve it. When it comes to an international issue, a hierarchist will feel inclined to favour a similar array of programmes and procedures. However, at the same time the hierarchist will also feel bound to the group that he or she represents (e.g. a particular state or ministry), and will tend to distrust representatives of other organizations involved in the international issue-area. As a consequence, a hierarchist will often engage in lengthy and formal international negotiations with a lot of emphasis on following the proper rules of international law and custom, as well as on the need to monitor the implementation of agreements. This reaction to the existence of international anarchy can seriously hamper the adoption of the extensive procedures and programmes favoured by the hierarchist. This tension may be called the *hierarchical dilemma in international relations*.

Max Weber has nicely captured the differences between the hierarchical and individualistic reactions to the existence of frontiers that I have tried to express (although he was obviously using different terminology):

> At the start two opposing attitudes towards the pursuit of gain exist in combination. Inside the community there is attachment to tradition and pietistic relations with fellow members of tribe, clan and household which exclude unrestricted quest for gain within the circle of those bound together by religious ties; externally absolutely unrestricted pursuit of gain is permitted, as every foreigner is an enemy to whom no ethical considerations apply. Thus the ethics of internal and external relations are completely distinct. The course of development involves on the one hand the bringing of calculation into the relations of traditional brotherhood, displacing the old religious relationship . . . At the same time, there is a tempering of the unrestricted quest for gain in relations with other foreigners.[42]

The ability of cultural theory to distinguish a limited set of quite different ways of reacting to anarchy is an important asset of the analysis. Other approaches either tend to assume that all international actors react in a similar way to the absence of world government (especially neorealism and neoliberalism – see Chapter 5)[43] or suggest an endless variety of different reactions to international anarchy (most notably constructivism and poststructuralism).

Three propositions

The eternal question is: so what? Why would we need a typology consisting of various international ways of life? First, because usually hundreds of people and organizations are involved in the decision-making processes of a transboundary issue-area. Without a set of ideal types, it would become very difficult to map the countless numbers of discrete opinions and actions that one comes across in empirical research, especially in case studies.[44] But the utility of a classification also reaches beyond the descriptive into the causal. A typology can also form the basis for attempts to find causal links and formulate general propositions. Various hypotheses can be derived from the model described above. Three are presented here.

> Proposition 1: *Change of an international regime will occur when the ways of life of the dominant organizations have come to be seen as incapable of providing explanations of, as well as solutions to, major developments and events within the international issue-area.*

The emergence of major social events that are inexplicable in terms of the cultural views held by powerful organizations may either empower other organizations within the issue-area or may lead the officials of those dominant organizations to change their cultural views.

> Proposition 2: *Cooperation within an international issue-area will be:*
> – *impossible to achieve when policy-makers are imbued with fatalistic thinking;*
> – *difficult to achieve when hierarchically oriented institutions dominate the issue-area;*
> – *easier to achieve between either individualistic or egalitarian organizations.*

For plural international regimes (i.e. regimes which allow hierarchical, individualistic and egalitarian organizations to all play an important part) the expectations are mixed. On the one hand, international negotiations in which egalitarian, individualistic and hierarchical organizations take part should be easier to conclude than negotiations dominated by hierarchical actors, as egalitarians and individualists tend to be more optimistic with regard to the feasibility of international cooperation. On the other hand, international cooperation is expected to be more difficult to achieve in plural regimes, as the inclusion of several ways of life will introduce rival conceptions of international issues and their possible solutions.[45]

This second proposition is largely based on the varying degrees of trust in foreign opponents that adherents to different cultural biases are expected to have.

Proposition 3: *Domestic and international institutions that allow a wide variety of organizations and citizens involved in an international issue-area to have access to the processes of problem-definition, information-gathering, and selection and implementation of policies will in the long-run lead to more successful solutions to transboundary problems than institutions that allow only part of these organizations to participate.*

In this hypothesis, the notion of 'a successful solution to a transboundary problem' is of course a highly contestable concept. This concept will be operationalized in the following manner. Only if adherents to all cultures collectively agree that an action or policy has been 'successful' will that action or policy be described as such. In this way it is hoped to avoid too extreme value judgements.

According to grid-group theory, each cultural bias has important contributions to make to the political debate. Each bias selects its own information and sees different risks and opportunities. Therefore, open, accessible decision-making processes are expected to be more resilient and to give rise to more creative thinking than decision-making processes that are built on views from only one or two cultural biases. Or at least this is the way in which the main policy precept of cultural theory has been presented thus far.[46] In the Netherlands, this advice has actually been taken up and implemented by the main governmental advisory council on health and environmental affairs (Rijksinstituut voor Volksgezondheid en Milieu

– RIVM).[47] However, the course of this research has resulted in a some-what different viewpoint, leading to a revision of proposition 3.

> Proposition 3 – revised: *Decision-making within an international issue-area is more likely to be successful if the relevant domestic and international institutions: (a) provide access to the decision-making processes to the followers of all ways of life; and (b) stimulate a dialogue and mutual understanding between the adherents to alternative rationalities.*

It is not only necessary that different cultural actors have access to the processes of decision-making. It is also vital that they are willing and able to listen to, and learn from, each other, otherwise a dialogue among the deaf will take place. Chapter 6 will make this clear.

Propositions 1 and 2 taken together constitute the claim that cultures matter in transboundary relations. They form the basis of the empirical analyses presented in Chapters 4 and 5, respectively. In these chapters it will be argued that the different ways of life followed by the actors involved in the protection of the Rhine river can explain a number of their actions and decisions. Proposition 3 specifies the claim that institutions matter as well in transboundary relations. This proposition is fleshed out more fully in the empirical research described in Chapter 6.

Part II
Cultures Matter

4
A Watershed on the Rhine

A puzzle

In 1979 the most severe diplomatic incident in the relations between the governments of France and the Netherlands since at least the Second World War took place when the Dutch ambassador in Paris was recalled by his government for consultation. The occasion for this unprecedented official protest was not (as might perhaps be expected) long-standing disagreements over the future shape of the European Community or large differences of opinion with regard to the combat of drugs or the need for a European nuclear force. Instead, the diplomatic row concerned the inability of the two governments to find an acceptable solution to a relatively minor environmental problem: the dumping of large amounts of salt into the river Rhine by an Alsatian mining company. This problem was only minor as it mildly affected very few interests: some Dutch water companies, the port of Rotterdam and a few horticultural firms. It did not threaten any flora or fauna in or along the river.

This unusual diplomatic incident was a manifestation of a general failure to achieve effective international cooperation on the protection of the natural environment of the Rhine between the riparian states of Switzerland, France, Germany and Holland, as well as Luxembourg (through which a large tributary runs) during the period 1945–87. In 1971 it was already widely recognized in the media, government departments and environmental groups that the Rhine had become extremely polluted – it was around this time the Rhine came to be known as 'the open sewer of Europe'. Two developments were responsible for the destruction of flora and fauna in the entire Rhine basin: (a) chemical and thermal pollution by industries, agricultural firms and communities along the river; and (b) the canalization of the Rhine. Up to 1987, despite extensive intergovernmental negotiations, it appeared impossible to agree on, and implement, any international measures to reverse the process of severe environmental degradation. Relations between the national delegations to the relevant international body, the International Commission for the Protection of the Rhine against Pollution (ICPR), were uncooperative and sometimes even distrustful. Environmental groups loudly condemned the lack of governmental action.

Since 1987, a strikingly different picture has emerged. At the end of 1996, the French newspaper *Le Monde* acclaimed the Rhine as 'the cleanest river in Europe'.[1] In the newspaper article, the international programmes to restore the ecology of the Rhine basin that have been developed since that year were partly credited with having cleaned up the river basin. This opinion reflects a much wider belief among all the organizations concerned with the protection of the Rhine that since 1987 intergovernmental cooperation on this issue has been extensive and exemplary. The relations between the members of the national delegations can only be called constructive, trusting and even amicable. Even the most radical environmental groups such as Greenpeace have praised the international agreements. At present, the European Commission sees the international environmental cooperation with regard to the Rhine as an excellent model for policy-making in other river basins.

This chapter will attempt to explain the sudden and remarkable transformation of the international cooperation on the ecology of the Rhine. For reasons of clarity, the discussion will be restricted to the changes in *intergovernmental relations* in this issue-area. This does not mean that the actions and opinions of non-governmental organizations will not be considered but that in this chapter only those actions and opinions of NGOs will be taken into account that have influenced intergovernmental relations. Chapters 5 and 6 will analyse other ways in which NGOs have affected the environmental restoration of the Rhine. One way of formulating the *puzzle* of this chapter is to ask: *what has caused the intergovernmental relations concerning the environmental protection of the Rhine to suddenly change from being limited, uncooperative and sometimes openly hostile to being extensive, effective and friendly?*

The puzzle can also be put in somewhat different terms. Until 1987, the scope and effectiveness of the international measures to protect the Rhine watershed were dwarfed by the extent and impact of domestic programmes for water protection. As will be shown later in the chapter, the clean-up of the Rhine that took place before 1987 was largely due to voluntary investments in water protection by industrial corporations along the river, as well as domestic water protection policies implemented by government departments. The clean-up of the river was certainly not driven by any international agreements. Since 1987, the situation has been reversed. From then on, the international agreements pertaining to the restoration of the Rhine catchment area have gone further than any domestic programmes for water protection. In fact, since 1987 the international Rhine agreements have led the way in the development of water protection policies, not only in the Rhine countries themselves, but also in the whole of the European Community. The puzzle of this chapter can therefore also be expressed thus: *how is it possible that before 1987 the international agreements concerning the environmental protection of the Rhine watershed were trailing domestic policies, whereas thereafter they have been leading the expansion of domestic water protection programmes in Western Europe?*

The chapter is structured as follows. First, the conceptualization of international regime change that has informed the case study will be presented. Elements of this conceptualization have already been discussed in Chapters 2 and 3, as it relies on the grid-group theory notion of 'surprise'. In the first section grid-group theory's depiction of policy change will be complemented with ideas formulated by Oran Young, Peter Hall and Paul Sabatier. Subsequently, the basic geographic features of the Rhine valley will be introduced as well as the main political institutions impinging on the international regime for the protection of the Rhine. Following that, a cultural analysis of the development of the Rhine regime since 1976 will be set out. The principles, norms, rules and procedures will be described that guided intergovernmental decision-making first before, and then after, 1987, and how these predispositions could change so quickly and fundamentally during that year will be explained. The analysis will be restricted to the period 1976–98, as it is too difficult to obtain reliable data on the mindsets and decisions of Rhine actors before that time. The concluding section of this chapter will reflect on what grid-group theory can contribute to our understanding of international policy change.

The rules of inference that have been followed in order to answer the puzzle of this chapter mainly consist of a comparison over time. To answer the question why the Rhine regime was fundamentally altered in 1987, not only will the events in that year be looked at, but also why the Rhine regime did *not* change in other years will be considered. More on this method of inference is contained in Chapter 1.

Change of international regimes

International change (including regime change) is notoriously difficult to explain. Sudden and positive international change is especially hard to account for. As set out in Chapter 2, grid-group theory offers several clues by which one can understand transboundary (as well as domestic) change more fully.[2] First, cultural theory often presents political parties and governments as coalitions of different ways of life. In doing so, it also highlights the potential sources of disagreement within parties and governments. Thus, it reveals dilemmas and tensions that may break up political parties and governments, or lead to a rupture in their policies. At the same time, cultural theory also points to issues about which organizations following alternative ways of life may find each other coming together. In this manner, by pointing out 'focal points' around which adherents to different cultures may coalesce, cultural theory also helps in understanding which new coalitions may be formed. These political transformations may have all kinds of transboundary effects.

In addition, there is the grid-group theory notion of 'surprise'. According to the framework, each way of life incorporates a number of truth claims, i.e. statements about how people, society and nature *really* are, and about how social problems can be solved. People are constantly comparing the validity of the truth claims of their own preferred ways of life to those of others. If a number of organizations have been dominant within an international issue-area for an extended period of time, it may happen that social problems emerge that can neither be understood nor solved with the help of their way of life. These problems are 'surprises' to the adherents of the dominant ways of life. When people begin to perceive these problems as huge and unacceptable, they will start to switch to other cultural views that seem to provide more satisfactory answers to the problems. The emergence of major social events that are inexplicable in terms of the cultural views held within powerful organizations may either empower other organizations within the issue-area or may lead the representatives of those dominant organizations to change their cultural views. In both cases, wholesale domestic and/or transboundary change will be the end result.

Stephen Krasner has distinguished between two forms of regime change.[3] The first, entitled change *within* a regime, involves solely the emergence of new rules and decision-making procedures in an international issue-area. The second form of regime transformation, called change *of* a regime, is more encompassing. Besides the adoption of alternative rules and procedures, it also includes the acceptance of new principles and norms within an international issue-area. The sudden and thorough transformation of the international cooperation to protect the Rhine that took place in 1987 was clearly a change of regime. Of the various ways in which cultural theory may account for international change, the surprise concept is the most promising candidate for explaining the remarkable overhaul of the Rhine regime. In 1997, another change of the Rhine regime took place. The new principles and norms adopted in that year even further strengthened the successful international cooperation on the revival of the Rhine that had taken off in 1987. This recent change of the Rhine regime cannot be explained with the surprise concept. It is better understood with the help of the 'shifting coalitions' idea – the first way in which grid-group theory can explain political change that was described above.

Conceptualization of transboundary change in terms of surprises is neither entirely new nor complete. Peter Hall, for one, has offered a similar picture of policy change.[4] According to Hall, decision-makers think in terms of 'policy paradigms', which contain not only specific policy positions, but also underlying theoretical and ontological assumptions. These policy paradigms will be abandoned only after the emergence of 'shocks', i.e. social problems that could neither have been foreseen nor solved within the reigning policy paradigms. Hall's shocks and cultural theory's surprises are clearly the same phenomena. As such, both accounts overlap. The account of policy change offered by Hall, however, includes elements that have hitherto been overlooked in cultural theory, and that complement its portrayal of policy and regime change. In particular, Hall argues that when a shock (or series of shocks) occurs within a specific issue-area, a whole new range of actors usually gets involved. In such a situation, decision-making is often lifted from the level of civil servants to the highest political level, the media starts to zoom in, citizens' groups become more active, and all kinds of experts offer their advice. A fundamental reversal of public policy seldom takes place without these developments. Politicians have more authority for making sweeping changes than do public officials. Extensive press coverage will increase the political salience of the issue. Citizens' groups and experts often defend policy paradigms that are quite different from those that have been followed. Without these factors, or so Hall argues, an overhaul of public policy is not likely. These factors (politicians taking an interest, media coverage, increased NGO activity, expert advice) are not mentioned in cultural theory, but are fully compatible with the theory's account of policy change. In my analysis of the course of the Rhine regime, I will show that these elements played an important role.

Oran Young and Gail Osherenko have also noted that external shocks or crises are important conditions for the adoption of environmental protection policies.[5] However, they have restricted their analysis to the impact that shocks have on the emergence of environmental protection programmes and procedures. They have tended to overlook changes in underlying principles and norms of environmental regimes. In other words, they have focused more on changes in a regime than transformations of a regime. Young's writings on international environmental protection include at least one element that should be added to the conceptualization of regime change presented here. Young has stressed the importance of 'entrepreneurial leadership' for the emergence of cooperation on ecological matters.[6] An entrepreneurial leader in international relations is someone who uses negotiating skills both to influence the ways in which transboundary issues are presented and to arrive at mutually acceptable solutions. Cultural theory has been rather silent on the role of leadership in policy change. Here again, the notion of entrepreneurial (and other forms of) leadership is highly compatible with the theory's portrayal of regime change. Once a major surprise has discredited the ways of life of dominant organizations in a regime, entrepreneurs will have a window of opportunity to get elements from their preferred ways of life instituted in the regime.

The list does not end here. Many other authors could be mentioned who have also pointed out that major policy reversals are often triggered by external shocks or surprises. One example would be Emanual Adler's views on cognitive evolution in international relations.[7] Another example would be the work of Paul Sabatier. There is a great likeness between Sabatier's 'advocacy coalition framework'[8] and grid-group theory's account of politics. Sabatier groups the manifold stakeholders in a policy subsystem (or domestic issue-area) into a few advocacy coalitions. Each advocacy coalition is held together by a common policy belief system, which consists of basic values, causal assumptions and problem perceptions. Each such coalition will busy itself with attempting to impose its preferences and perceptions on the other stakeholders in the policy subsystem, using its financial resources, expertise, legal authority and so on. This comes very close to cultural theory's struggle between ways of life. Besides external shocks, Sabatier also emphasizes the role of policy brokers in effectuating radical policy change. This element should definitely be taken up in the conceptualization of regime change used in this chapter.

The overlaps that sometimes exist between cultural theory and other frameworks, as well as the compatibility of cultural theory with other approaches, should not be seen as problems. In my view, if anything, they are assets of grid-group analysis. This is all the more so, because cultural theory may help to explain certain social processes that other frameworks cannot. With regard to understanding international regime change, cultural theory may also have something extra to offer. Compared to Hall's policy paradigms and Sabatier's belief systems, the four ways of life of cultural theory are much more specific. The ways of life include particular preferences for social relations, risk attitudes, views of human nature, etc. As the ways of life are spelled out more, they offer more guidance and support for a researcher interested in explaining international regime change. This is one advantage of grid-group theory.[9] Or at least it would be if the four ways of life were suitable ideal types for describing the actions and motivations of real actors. But that is an empirical question, not a theoretical one.

A further advantage is that grid-group theory sheds some more light on the *direction* in which an international regime is likely to change after a shock has occurred. Because the ways of life distinguished in cultural theory are more specific and comprehensive than the policy paradigms of Hall or the belief systems of Sabatier, grid-group theory is somewhat better suited to explain which actors are empowered by a certain shock and which actors are not. Those actors who manage to portray their way of life as offering a better response to the surprise will be helped by it. Other actors whose way of life comes to be seen as inadequate to deal with the shock will be discredited.

A last advantage of grid-group theory is that the analysis is somewhat better suited to explain why certain shocks have *not* led to fundamental change. Not every surprise that occurs in an international issue-area will lead to an overhaul of the international regime regulating that issue-area. As grid-group theory spells out its ways of life somewhat more a priori than comparable approaches, it is less likely to offer circular explanations of why a surprising event did not lead to large-scale change. (It can be expected that other, more inductive, approaches would say that the shock was not 'shocking enough'.) The history of the Rhine regime will provide an opportunity to exemplify this line of reasoning.

It could be argued that the conceptualization of wholescale change of international regimes that is offered here does not add up to much. It could, for instance, be faulted for using exogenous shocks as explanatory factors. Also, it does not seem to go very far beyond a commonsense, intuitive understanding of regime change. But (almost) everything is relative, and this conceptualization is believed to fare better than what other IR approaches have to offer. In the account of the transformation of the international regime for the protection of the Rhine that follows below, several developments will be pointed out that various other frameworks could not have explained.

But first we look at the Rhine.

A view of the Rhine

The Rhine begins in the Swiss and Austrian Alps, where a number of small brooks flow together in the Bodensee. From this lake, the waters of the Rhine start their 1320 km trip to the North Sea through Switzerland, France, Germany and the Netherlands. The first section of the river partly forms the German–Swiss frontier. Further downstream, between Basle and Karlsruhe, the Rhine forms the border between France and Germany. In the Netherlands, the Rhine scatters over the whole country, supplying all but the most northern parts with water. Its main artery, however, flows through the city of Rotterdam. Other countries included in the Rhine catchment area are Belgium, Luxembourg, Austria and Italy. These countries are connected with the Rhine by tributaries, such as the Moselle, which is an important tributary that flows through Luxembourg. Since 1996, the Rhine–Main–Donau canal in Bayern (Bavaria) has been in use, linking the river with Central Europe.

For many centuries, the transport of bulk goods over the Rhine has been of paramount importance to the economies of Western Europe. One natural characteristic has made this extensive shipping on the river possible. The Rhine has a massive and steady flow of water. In the warm months of the year, snow in the Saint Gothard massif melts and drains into the Rhine. In the cold months, rainfall all over the Rhine basin makes up for the reduced inflow from the Alps. This steady flow of water has stimulated trade and shipping for many centuries. Before the nineteenth century, commercial activities were hampered by the tolls levied by robber-barons situated along the waterway, as well as by existing rock formations within the river and by the meandering and braiding of the river. In 1815 the Peace Conference of Vienna established the principle of free navigation of the Rhine, which was further strengthened by the 1868 Navigation Act of Mannheim. At the Vienna Conference, the Central Commission for the Navigation of the Rhine (CCNR) was also founded. Ever since, its tasks have been to ensure free navigation and to promote projects (such as canalization) that increase the possibilities for shipping.

Shipping on the Rhine has heavily benefited from, as well as contributed to, the massive industrialization and urbanization that have taken place along its shores. The river connects the biggest seaport in the world (Rotterdam) with the largest inland port on the globe (Duisburg). The German Ruhr area remains one of the biggest centres for steel production in the world. About 20 per cent of the world's chemical industry is situated along the banks of the river. Other important industries consist of paper and pulp companies, as well as potassium and coal mines. At present, the river receives the wastewater of about fifty million people, chemical and other industries, potassium and coal mines, as well as agricultural sources. It is also used as a source for drinking water for more than twenty million people.

The canalization of the Rhine, as well as the discharge of polluted wastewater from industrial, agricultural and municipal sources, has had a devastating impact on the ecosystems of the river, at least until the middle of the 1980s. Records from both governmental and non-governmental organizations show that the environmental degradation was at its peak in the late 1960s and early 1970s. The events that took place between 8 and 10 June 1971 are particularly revealing. During these few days, a long stretch of the waterway (the 100 kilometres between the cities of Mainz and Koblenz) contained literally no oxygen. This absence of oxygen was caused by extreme environmental degradation as well as adverse weather conditions. Not surprisingly, in the section between Mainz and Koblenz several thousand tonnes of fish died and covered the river surface from shore to shore for about 48 hours.[10] The Rhine came to be known as the 'open sewer of Europe'.

Disregarding the ineffective Salmon Treaty of 1885, the first serious attempts to achieve international cooperation on the protection of the Rhine started in 1946 when the Dutch delegation to the CCNR started to raise environmental issues.[11] Four years later, Switzerland, France, Germany, Luxembourg and the Netherlands formed the International Commission for the Protection of the Rhine against Pollution (ICPR). In 1963, the ICPR acquired official status with the signing of the Bern Convention on the Protection of the Rhine. In 1976, the European Community (EC) also became a party to the Bern Convention. Ever since its official creation, the tasks of the ICPR have been (a) to report on the state of the environment of the Rhine; (b) to propose international policy solutions to the Rhine's ecological problems; (c) to hold regular international consultations; and (d) to monitor and partly implement any intergovernmental agreements that have been reached.

The ICPR is organized on the following basis. It has a small secretariat in Koblenz, Germany, which provides administrative support. The Commission itself consists of high-level civil servants from the riparian countries and the EC, who usually work for the ministries that have lead responsibility for water protection within their respective countries. Sometimes, however, they have been diplomats from the various Ministries of Foreign Affairs. The civil servants who make up the ICPR meet annually, bringing a national delegation to these plenary meetings made up of lower level officials from their own ministries as well as public servants from other ministries. At these meetings, work programmes, finances and formal procedures are settled. A coordination group, convening four times a year, is responsible for the actual planning and coordination. The presidency of the ICPR rotates among its members, although the executive secretary has always been Dutch. Possible solutions to specific issues (such as how to prevent floodings) are prepared and discussed in working groups and expert groups, each such group being made up of national governmental experts. The ICPR receives its instructions through the Ministerial Rhine Conferences that are held every two to three years. At these conferences, the responsible ministers formulate the political goals of the Commission, and evaluate ongoing and completed activities. In 1976, the ICPR was able to decide upon the first international agreements. In that year, both the *Convention on the Protection of the Rhine against Chemical*

Pollution and the *Convention on the Protection of the Rhine against Chlorides* were signed.[12] Another task undertaken by the International Commission for the Protection of the Rhine has been the installation of an extensive system for measuring the water quality of the river. Another finely tuned measurement system has been operated by a non-governmental organization, the Internationale Arbeitergemeinschaft der Wasserwerke in Rheineinzugsgebiet (IAWR). This is a transnational organization representing the drinking water companies in the Rhine basin. The results of their measurements are published in the yearly reports of the IAWR.

A history of the international regime to protect the Rhine is presented below in terms of the cultural approach that developed in this and the previous chapter.[13] As mentioned above, this history will be limited to the period 1976–98, which covers the most important international policy changes to date. Furthermore, only the changes in the principles, norms, rules and programmes that were adopted, or followed, in the context of the ICPR negotiations will be discussed. In other words, only the intergovernmental principles, norms, rules and procedures concerning the protection of the Rhine's environment will be considered.

To reconstruct the history of the Rhine regime official publications from the involved organizations, a few internal documents and the academic literature on the topic were used. In addition, 54 semi-open interviews were held with 58 stakeholders during 1996 and 1997 (see Appendix A). These interviews were meant to tease out not only the international policy processes (what was decided or proposed by whom and when), but also the cultural backgrounds of representatives from the organizations.

1976–86: a hierarchical international regime

The years 1976–86 were mainly taken up by efforts to ratify, and implement, the Chemical and Chlorides Conventions signed in 1976, as well as by attempts to agree on a convention against thermal pollution of the Rhine. The principles, norms and decision-making procedures that the governmental delegations followed in their efforts to agree on these treaties were by and large hierarchical. Before this is set out in detail, it may be helpful to briefly recapitulate the hierarchical approach to international regimes (see also Chapter 3). Hierarchists put their faith in expert knowledge (as expert knowledge provides an excellent justification for a stratified society). Therefore, when faced with a domestic problem, a hierarchist will propose a whole battery of sophisticated methods and programmes with which to study and measure the problem, as well as solve it. When it comes to an international issue, a hierarchist will feel inclined to favour a similar array of programmes and procedures. However, the hierarchist will also feel bound to the group that he or she represents (e.g. a particular state or ministry), and will tend to distrust representatives of other organizations involved in the international issue-area. As a consequence, a hierarchist will often engage in complex and formal international negotiations with a lot of emphasis on following the proper rules of international law and custom, as well as on how to monitor the implementation of agreements. This reaction to the existence of international anarchy (which can be defined, notably, as the *absence of formal hierarchy within the world system*) will seriously hamper the adoption of the extensive procedures and programmes proposed by the hierarchist. This might be called the 'hierarchical dilemma in international relations'.

This dilemma is well brought out by the development of the international regime to protect the Rhine during the period 1976–86. During this time, both the Chemical and the Chlorides Conventions were not implemented, while the efforts of the ICPR to negotiate a convention on thermal pollution came to naught. These failures were accompanied by much acrimony between the governments of the Rhine countries. In the meantime, however, the same governments developed extensive water protection programmes within their own countries. Moreover, the corporations along the river made a number of voluntary investments in water protection. The annual reports of the IAWR (the umbrella organization of the water supply companies) clearly show that these domestic measures by governments and industries led to a significant decrease in the pollution of the Rhine. In fact, in 1986 the levels of many toxic substances in the waters of the Rhine were already down by some 80 per cent (compared to 1970).[14] Still, this reduction of pollution was not matched in any way by a decrease in the bickering among the national delegations to the ICPR. The hierarchical principles, norms, rules and procedures that prevailed in the Rhine regime were at the root of this odd situation. Below, the principles, norms, rules and procedures that made up the Rhine regime during 1976–86 are first described, and then how these regime elements can explain the simultaneous existence of successful domestic water policies and antagonistic international negotiations will be discussed.

Principles and norms

A first principle (or norm) has been defined as 'the degree to which actors trust each other, and are willing to engage in international cooperation'. The period from 1976 until the end of 1986 was characterized by low levels of intergovernmental trust and cooperation. All three attempts to implement international treaties during 1976–86 show this. The aim of the Rhine Chemical Convention was to reduce pollution of the river by discharges of hazardous chemical substances (including heavy metals) from chemical plants, community sewage systems and agricultural lands. The means by which these goals were to be achieved consisted of the formation of a black and a grey list of toxic chemical substances that were going to be regulated. The black list was to contain the most toxic chemical substances whose reduction needed priority. The grey list was to include substances that were judged less hazardous but still in need of regulation. Black-listed substances were to be dealt with at the international level. The national delegations to the ICPR were to achieve unanimity on emission standards for these materials, based on the principle of 'best technical means'. Regulation of the grey-listed substances was left to national discretion.

The process of implementing the Chemical Convention was beset with difficulties.[15] The ICPR first drew up a list of 83 substances falling within the scope of the black list. Under the rules of the Convention, the national representatives in the ICPR were to agree on the *effluent limits*[16] for the black-listed chemical substances, a process which never took off. The delegations wanted to be certain that their own national interests (as they perceived them to be) were no less served than those of the other states. In particular, the German representatives insisted that German industry should not be burdened with environmental obligations not prevalent in other countries. The German delegation blocked the adoption of effluent limits for all chemical substances that had not been regulated across the whole of the European Community.[17] As a result, in the period 1976–86, the ICPR was able to formulate concrete emission standards for just three black-listed substances (quicksilver, cadmium and tetrachloride), and two of these three effluent limits were only adopted after similar EC directives had been issued.[18] The insignificance of this number becomes apparent on the realization that the European Commission had drawn up a list of some 1500 chemical substances suspected to be toxic.

Low levels of trust and cooperation were also characteristic of the implementation of the Rhine Chlorides (or Salt) Treaty.[19] Before 1976 an average load of 400 kilograms per second (kg/s) of salt were being dumped into the river. An Alsatian mine company, Mines de Potasse d'Alsace, was responsible for about 35 to 40 per cent of this load. The salt emissions were especially harmful to the interests of several Dutch water companies, flower growers and the port of Rotterdam, though they did not significantly affect the flora and fauna of the river. The Salt Treaty of 1976 focused on the emissions of the Mines de Potasse d'Alsace. According to the treaty, the emissions of the mine company were to be cut by 60 kg/s in three phases, and the salt not dumped into the river would be injected deep into the Alsatian earth. The costs were estimated at 132 million French francs. The Netherlands would finance 34 per cent of the project, Germany and France would each pay 30 per cent of the costs and Switzerland the remaining 6 per cent. For a number of years, the French government refused to put the treaty before parliament for ratification. It argued that the opposition to the treaty organized by Alsatian members of parliament was too strong and instead kept proposing to study new sites for salt storage. The Dutch government (and media) did not believe that the French government was sincerely committed to finding a solution to the salt issue, and as a result there was much acrimony. In 1979 the Dutch government took the remarkable decision to recall its ambassador from Paris for consultation. Even today both Dutch and French officials who were involved in the salt negotiations some ten to twenty years ago, are still suspicious of each other's motives at the time.[20] During one interview, a former member of the Dutch delegation to the ICPR even used the word 'lying' to describe the activities of the other side – a word almost anathema in diplomatic jargon.[21] The cities of Rotterdam and Amsterdam, as well as several Dutch horticulturists, lost faith in the ability of the governments to solve the salt problem and started a series of remarkable court cases under private international law in both France and the Netherlands,[22] which further increased the tensions between the two governments. However, in 1983, the new Mitterrand government finally succeeded in ratifying the treaty, and two years later the first phase of the treaty was executed, which consisted of a 6 per cent reduction of the average salt load passing the Dutch–German border. The other two phases foreseen by the treaty were never to be implemented.

The failed efforts to achieve international cooperation on the two

Rhine conventions were complemented by extensive, yet totally ineffective, attempts to agree on another treaty. Until 1987, ICPR officials spent years preparing and designing an international treaty to prevent thermal pollution (i.e. heating) of the Rhine. In the end, the national delegations simply could not agree upon its content.

The low levels of trust and cooperation were not limited to the relations among the national delegations. The representatives of the European Commission were constantly arguing with national delegates over who had the right to negotiate over which subjects with the only non-EC party to the Rhine Convention, Switzerland.[23] Also within the Deutsche Kommission zur Reinhaltung des Rheins[24] there was an ongoing struggle over jurisdiction and competence, especially between the *Länder* and the federal ministries.

In sum, the history of the (non-)implementation of the Rhine Conventions shows that strife and distrust plagued the international efforts to reduce chemical pollution of the river between 1976 and 1986. This characterizes the hierarchical attitude towards international relations.

A second principle/norm has been specified as a preference for a particular form of international governance. During the period 1976–86, the Rhine regime was governed according to hierarchical ideals. First, the international consultations on the Rhine were strictly intergovernmental. There were virtually no formal or informal contacts between Dutch, French, German and Swiss environmental groups[25] on the one hand, and the ICPR on the other. Nor were firms consulted or heard. Furthermore, the delegations to the Rhine meetings always followed a strictly formal route, scrupulously adhering to the established traditions of intergovernmental negotiations and international public law. Proposals on international action were initiated at the Ministerial Rhine Conferences. These proposals were then discussed, developed and revised in the working groups of the ICPR among the lower ranking members of the national delegations. Minor policy proposals were passed on to the plenary meeting of the ICPR. Major policy initiatives were put on the agenda of the following Ministerial Conference. In the latter case, the proposals would be moulded into international conventions or appendices to such treaties. If unanimous agreement at the ministerial level was possible, conventions would be solemnly signed and the process of ratification would start. Only after ratification by national parliaments and the inclusion of the convention in domestic law would the convention enter into force. This meticulous following of the traditions of public international law is a quintessential trait of the hierarchical way of life.

Rules and procedures

Rules are specific prescriptions or proscriptions for action, while procedures are prevailing practices for implementing collective choice. Before 1987, few rules and procedures were actually adopted in the Rhine regime, due to the unwillingness of the national delegations to cooperate. However, as the above has shown, many (largely fruit-less) efforts were undertaken to establish rules and procedures. These proposed rules and procedures came close to the hierarchical ideal of managing environmental issues. In essence, the ICPR delegates tried to construct a 'command-and-control' approach to the environ-mental protection of the Rhine. Such approaches to environmental issues were applied domestically by many Western governments at the time, and the ICPR members wanted to instal such an approach at the international level as well. They attempted to lay down effluent limits for a number of black-listed chemical substances. These effluent limits were to be legally binding in all riparian countries, and were to be based on the toxicity of the chemical substances as well as on the wastewater technology available to firms. This detailed, top-down approach is a hierarchical way of solving environmental problems.

Above, it has been shown that it was well-nigh impossible to achieve international coordination on the protection of the Rhine before 1987. There is of course not much new to the conclusion that international cooperation is difficult to achieve, especially be-tween upstream and downstream states. It is in line with both realist and game-theoretical approaches. What these frameworks would not be able to explain adequately is the emergence of extensive and overlapping domestic policies concerning an environmental issue on which international coordination has stalled. From the 1960s onwards, all Rhine states developed legal systems to protect the natural condition of their waters.[26] Under these systems, permis-sion to drain into rivers and lakes was only given to communities and firms if their discharges of wastewater met a number of effluent standards, and if the open waters into which they led their emissions met certain water quality standards. Moreover, municipalities and companies were required to apply a minimum level of technology in the treatment of their wastewater.[27] These domestic policies, in tandem with voluntary measures taken by industries, worked quite well. As stated above, from the early 1970s onwards the presence of most suspected toxics started to decline rapidly. It is quite paradoxical that the delegations to the Rhine conferences could not agree on international measures, when in their own countries great public and private efforts were being made to reduce the pollution of the Rhine.[28] Realist and game-theoretic models would only be able to explain this paradox partially, if at all. These approaches would highlight the possibility of free-riding, unequal relative gains and cheating as factors that plague and hinder international regulation more than domestic regulation. However, it could be argued that exactly these factors were much reduced in the case of the Rhine by the development of national water policies. The cultural approach developed here is able to offer a fuller account – in terms of the hierarchical principles, norms, rules and procedures that dominated the Rhine regime between 1976 and 1986.

The cultural explanation would start with the lack of trust that existed among the ICPR delegates (the first hierarchical principle/ norm described above). There is a problem of inference here. In part, the lack of trust and cooperation among ICPR members is to be derived from the meagre results of their international negotiations. It then becomes quite difficult to explain the lack of international coordination on the basis of the absence of trust among ICPR members as this would look like circular reasoning. One counterargument, however, can break the circle. Not every failed attempt at international cooperation is necessarily accompanied by so much friction as were the Rhine negotiations before 1987. The sour relations between France and the Netherlands over the relatively unimportant salt issue (resulting in the recall of the Dutch ambassador from Paris) are a particular case in point. The antagonisms present in the Rhine deliberations seemed to have gone further than the normal disappointment that may accompany a failed attempt at international coordination. Therefore, it seems that it *is* possible to argue that the Rhine negotiations were dominated by distrust among the delegates, and that this distrust prevented effective international coordination on the protection of the Rhine, despite extensive and overlapping domestic policies to clean up the river. From the interviews it appeared that this view also corresponds with the opinions of the persons who were involved in the negotiations at the time, these officials often citing 'lack of trust among the parties' as the main cause of the deadlock.

A further element of my cultural explanation consists of the second hierarchical principle/norm described above, namely the preference for a highly formal, legalistic way of intergovernmental decision-making among ICPR delegates. This preference for the adoption of formal international conventions tended to stall international decision-making concerning the Rhine as well. For one thing, the signing and ratification of official international Rhine treaties necessitated endorsement of these initiatives in many fora. The Rhine conventions not only had to be unanimously supported by all involved governments, but also required the official blessing of the parliaments in each riparian country as well as the European Community. This took up a lot of time and gave opponents many opportunities to block the adoption of the international agreements. Moreover, as both André Nollkaemper and members of the ICPR secretariat have suggested,[29] the governments of the riparian countries seem to have been wary of formally committing themselves to agreements legally binding under international public law. International conventions create a legal obligation for governments to take certain measures, and the Rhine states did not seem comfortable with taking up such formal obligations. Yet, at the same time, this was exactly the way in which the delegations tried to proceed.

The hierarchical rules and procedures that the delegations were attempting to create for the protection of the Rhine provide a third, complementary explanation. As noted before, the governments involved in the Rhine meetings aimed at developing an intergovernmental command-and-control system for water protection. It could be argued that such a top-down approach is not very compatible with the existence of command-and-control systems at the domestic level, yet this was the case in the Rhine area. In each riparian country, command-and-control policies for environmental protection were being put in place. The basics of these policies were the same in all countries: each country regulated point source emissions by issuing discharge permits that were subject to effluent limits; every point source polluter in the riparian countries had to pay a fee to the government, the amount of which was dependent on the amount of pollution; non-point discharges (i.e. discharges by farmers) were regulated through bans on the use of specific pesticides as well as less drastic means; and in each riparian country, water quality standards were formulated. Yet despite all these basic commonalties between the water protection policies of the Rhine governments, many details oscillated wildly. In the Netherlands, only a few effluent guidelines were adopted. The issuing of pollution permits to firms and cities was mostly based on a number of water quality standards and the civil servants responsible for granting permits used these water quality standards as guidelines for the effluent requirements they imposed on individual point source dischargers. In the process, they also took into account the financial circumstances of polluters.[30] The German *Länder* developed many more effluent limits than the Dutch authorities which were specified for each branch of industry.[31] In France, a very complex system for water protection was set up. Inventories were made of the environmental conditions of all river and lake basins in France and the expected future human use of each river or lake basin was also assessed. In addition, the Ministry of the Environment (in tandem with other ministries) published a number of general water quality standards. On the basis of all of these factors, *cartes départementales d'objectives de qualité* were drawn up, stating the environmental goals for each area of surface water in each *département*. It was up to the *préfets* to reach these goals in their *départements* by the issuing of permits – in cooperation with officials from the Ministry of the Environment and the regional *agence de l'eau*.[32] In Switzerland a much simpler system was set up, based on 52 general effluent limits and a number of water quality standards adopted by the federal government in 1975.[33] This listing of various differences in the national water protection policies of the riparian countries does not do full justice to the many intricate ways in which these systems were set up, functioned and differed from each other. However, they illustrate the main point, that the water protection systems set up by the authorities in the Rhine countries were all founded on a command-and-control logic, but at the same time differed in many ways. Given this, the attempts to set up another command-and-control approach at the intergovernmental level was bound to fail. Such an intricate top-down approach was simply not compatible with the diverse water protection systems that existed at the domestic level.[34]

In conclusion, hierarchical principles, norms, rules and procedures dominated the Rhine regime between 1976 and 1986. This helps to explain why the international negotiations regarding the restoration of the Rhine were completely stuck in 1986, even though domestic policies and private investments had already achieved a significant clean-up of the river by then. In the summer of 1986, top officials from the Dutch Ministry of Transport, Public Works and Water Management (in Holland the lead agency for international environmental cooperation concerning water basins) tried to set up a meeting with their counterparts in the German government to discuss the protection of the Rhine once again. The German public servants refused to meet the Dutch officials.[35] This serves as an illustration of how bad the intergovernmental relations on the protection of the Rhine had become by the middle of 1986.

1987–95: an international regime based on two ways of life

Below, the process that changed the Rhine regime is first described, then the elements of the new regime are outlined.

Sources of change

In late 1986 an accident took place that triggered a fundamental overhaul of the hierarchical international regime to protect the Rhine.[36] On the night of 1 November, a fire broke out in warehouse 956 of Sandoz AG in Basle, Switzerland. The warehouse contained 1351 tonnes of various chemical substances, some of which were extremely toxic. Firefighters worked throughout the night to put the fire out, while civil defence sirens wailed in the city of Basle and police used loudspeakers to warn people to stay inside and protect themselves against the fumes. The fire brigades sprayed millions of litres of water on the warehouse in their attempts to put out the fire. This amount of water was too much for the catch basins to hold, and in the morning about 10 000 to 15 000 cubic metres of water, mixed with highly dangerous chemicals, had flowed into the bordering Rhine. Basle was shrouded in (as it turned out later) relatively harmless but intimidating red clouds. The chemicals that washed into the Rhine formed a red trail 70 kilometres long moving downstream from Switzerland, to France, Germany and the Netherlands at 3.7 kilometres per hour. Hundreds of thousands of dead fish and waterfowl washed up along the banks of the river. A number of sheep grazing along the Alsatian side of the river died. All the processing plants of the drinking water companies in Switzerland, France, Germany and the Netherlands that were using Rhine water were shut down.

As 48 out of 58 interviewees agree, and no one challenges, the immediate impact of the Sandoz accident was to discredit the existing practices to protect the environment of the Rhine.[37] The mass media contributed to this by giving extensive coverage of the spillage and

its effects, as well as by blaming the governments of riparian countries for the debacle. Newspapers and journals in both Europe and
the United States used headlines such as 'Europeans do it to each
other', 'Bhopal on the Rhine', and 'Wir sollten aufwachen und
überlegen'.[38] In Basle and other cities on the Rhine, large demonstrations were organized by environmental groups. The ministers
responsible quickly convened.

In terms of the cultural approach used here, the Sandoz incident
shook the belief in a strictly hierarchical approach to the environmental protection of the Rhine and showed that the international
delegates could no longer persevere in their formal and fruitless
attempts to clean up the river. Also, as Peter Hall would have predicted, the Sandoz incident motivated actors (such as journalists
and ministers) to pay much more attention to the issue of the river's
pollution. These developments made room for the international
regime to be infused with other ways of thinking, and it can be
shown that since 1987 the international regime for the Rhine has
become a hugely successful combination of individualistic principles
and norms with hierarchical rules and programmes.[39] This fruitful
combination was possible because of the complementary strengths
of both ways of life. As has been illustrated above, the hierarchical
way of life faces a dilemma when applied to international relations. On the one hand, it calls for the execution of a whole battery
of international solutions to transnational issues. On the other, it
tends to make international negotiations difficult by insisting on
formalistic negotiations, and by jealously guarding perceived national
interests. The individualistic way of life faces an inverse problem,
at least when applied to international environmental issues. If convinced of the seriousness of international environmental degradation,
an individualist would favour a quick and pragmatic fix to ecological problems. However, cultural theory assumes that an individualist
tends to perceive environmental degradation later than others. The
Sandoz incident destroyed both the hierarchist's rationale for lengthy,
formal negotiations and the individualist's tendency to deny the
acuteness of environmental degradation. This, in the abstract, made
possible an effective mix of the two ways of life. The resulting regime
contained individualistic principles and norms, but hierarchical rules.

However, to paraphrase Thomas Risse, 'ideas do not mix freely',
but have to be proposed and combined by people and organizations.[40]
This raises the questions: (a) who were the persons advocating
individualistic ideas? and (b) why were they more influential than

the people and organizations favouring egalitarian alternatives? An important source of change was the Dutch Ministry of Transport, Public Works and Water Management, which headed the national delegation of the Netherlands to the Rhine conferences. For many years, the Ministry had been led by Ms Neelie Kroes, a prominent member of the People's Party for Freedom and Democracy (VVD). Of all the major political organizations in the Netherlands, this party has come closest to the individualistic ideals of Ronald Reagan and Margaret Thatcher: less government regulation, lower taxes, more allocation through markets and more private initiative and responsibility. The Sandoz disaster changed the views of Minister Kroes in two ways.[41] First, it made her realize how pressing the ecological problems of the Rhine basin had become. This is a clear example of someone thinking along individualistic lines, who can no longer deny accumulating evidence of an environmental issue. Second, it made her aware that the salt question was not the most important environmental issue of the river, but was much more a financial problem for some water supply companies and horticul- turists. As a consequence her attention turned away from the salt problematic towards the pollution of the river by chemicals and heavy metals, as well as the destruction of fauna and flora by the construction of dikes and dams. Thus, the intractable chlorides issue more or less disappeared from the international agenda, opening the door to intergovernmental cooperation on other issues. To achieve this cooperation in a swift way, Minister Kroes took a step that is highly unusual in international relations, yet typically individual- istic. To broker international agreement on the clean-up of the Rhine she relied on private initiative. She hired a team of consultants from McKinsey-Amsterdam to outline a comprehensive international agreement on the restoration of the Rhine basin and to build up the necessary intergovernmental support for this plan. In other circles, this was frowned upon. A public servant doubted the wisdom and ethics of 'letting private organizations perform governmental ser- vices',[42] while an environmental group criticized the decision for 'using an old-boy network'.[43] This was because one of the members of the McKinsey team was Dr Pieter Winsemius, Dutch Minister of the Environment from 1982 to 1986 and another prominent mem- ber of the VVD, the political party of Ms Kroes that tends to support individualistic policies.

The McKinsey project team developed a strategy to set up effective international cooperation. This strategy contained three important new elements. First, the consultants held interviews with the scientists who, according to their own governments, were the top two experts on water pollution within their countries. Each of these scientists was asked which substances deserved the highest priority in the clean-up of the Rhine. The overlap of the answers formed an undisputed list of substances that had to be kept out of the river. Another element of the McKinsey plan was the introduction of a single quality aim which could act as a symbol for the restoration of the Rhine and which would stand out in people's minds and generate positive feelings. This quality aim became the return of salmon to the Rhine by the year 2000. The rationale behind this choice was that salmon form the top of the food chain in the river, and their return therefore necessitated a whole array of environmental matters. Both the terms 'salmon' and '2000' were thought to create sympathy and visibility for the restoration plan. The last point of the McKinsey plan was to keep international regulation informal and to a minimum – a typically individualistic norm for international governance. Agreements were not to be laid down in official international treaties but in non-binding reports issued by the ICPR to which the delegations would pledge themselves only verbally. Also, the international agreements would be restricted to the *goals* of environmental restoration. How these goals were to be achieved was left to each national or local authority. In this manner, the competency struggles between the German federal ministries and the *Länder*, as well as between the EC and the national representatives, could be sidestepped. The McKinsey team travelled to all the capitals of the Rhine countries, building support for their plans, talking to ministers and civil servants, winning them over. In the summer of 1987, their report was endorsed by the authorities of the riparian countries. It formed the basis of the ambitious 'Rhine Action Programme' (RAP), to which the ministers of the countries concerned committed themselves at the end of 1987, just one year after the Sandoz fire.

The involvement of Minister Kroes and the consultants from McKinsey explains the infusion of the Rhine regime with the individualistic principles and norms which are spelled out below. It does not provide an explanation for why egalitarian notions and practices were not also taken up. Such an explanation is called for as the egalitarian discourse, full of stories of how multinationals are destroying the natural environment, seemed very much validated by the Sandoz disaster. Again, as Hall, Sabatier and Young would have expected, the explanation must be found in the actors advocating these ideas. At the time, there were no representatives of egalitarian attitudes towards the environment at the ministerial level. Such environmental groups as Greenpeace, Reinwater and Bundesverband Bürgerinitiativen Umweltschutz did come close to the egalitarian ideal type. However, the radical actions and language that a few environmental groups undertook and used after the Sandoz fire[44] antagonized government officials.

Principles and norms

The first principle/norm distinguished in Chapter 3 concerns the degree of trust that actors have in international cooperation. It was also argued that individualistic approaches to international cooperation are characterized by a relatively large degree of trust in the willingness of other actors to keep their promises and honour their contracts. (It is very difficult to justify social relations that are free from restrictions on personal liberty, if this kind of trust is lacking.) After the Sandoz affair, international cooperation and trust suddenly abounded in the Rhine regime. In less than one year, unanimous agreement was reached on the ambitious Rhine Action Programme (RAP) on 1 October 1987.[45] The goals of the Rhine Action Programme were threefold: (a) to enable the return of salmon, and other higher species, to the river by the year 2000; (b) to ensure continued use of the Rhine as a supply of drinking water; (c) to clean up the sediment in the river that had become polluted by heavy metals and chemical substances. These aims may perhaps look quite ordinary but were in fact far-reaching, necessitating the achievement of many subgoals. The continued use of the Rhine for drinking water demanded reductions of chemical substances from both point and diffuse sources. It was decided to reduce the discharges of priority substances by 50 per cent by the year 1995. The return of salmon, as well as other species at the top of the food chain, not only made reduction of discharges necessary, but also required a number of hydrological and morphological changes to the river. Salmon and other fish used to hatch in upstream spawning grounds in Germany and France. Their offspring would swim down the river to the North Sea, and after several years they would return to their birthplaces to reproduce themselves. However, in the course of the twentieth century these upstream spawning grounds had either disappeared altogether because of industrial, agricultural or city development, or had become unreachable because of the presence of dams and weirs in the river. The reintroduction of salmon therefore entailed the reconstruction of spawning grounds at the expense of human usage of territory, as well as the building of fish passages around the dams and weirs. The RAP was not only comprehensive, but also costly. It has been estimated that the German chemical industry along the Rhine by itself spent DM 6.6 billion on sewage treatment during the period 1987–91.[46] The costs incurred by the communities along the river are generally thought to be higher. In addition, a total amount of DM 110 million was foreseen for the construction of fish ladders and fish sluices.[47]

A further characteristic of the remarkable cooperation on the environmental protection of the Rhine was the timely implementation of the Rhine Action Programme. Originally, the goals of the RAP had to be achieved by the year 2000. By the end of 1994 most of the aims had already been reached. Discharges of almost all priority substances had been reduced by 70 per cent (instead of the targeted 50 per cent).[48] The salmon and other species had also returned to the river, for the first time in forty years.[49] A sophisticated warning system had been put in place to provide a more adequate reaction to accidental spillages.[50] Therefore, as early as 1995, work could start on the drafting of a new international agreement (the Convention for the Protection of the Rhine) that was adopted by the responsible ministers in January 1998. This new agreement focuses on the remaining environmental problems, including the continuing discharges of chemical substances (especially nutrients) from diffuse point sources, and the removal of remaining polluted sediment. It also emphasizes the ecosystem approach to the protection of the Rhine basin.

The European Commission also contributed to the cooperation. Whereas before 'Sandoz' the Commission had been constantly involved in competency struggles with the national delegations, it now voluntarily withdrew itself from the negotiations and contributed to the financing of the Rhine Action Programme. In 1991, even the seemingly unsolvable salt issue could be put to rest, after a continuous struggle of about 45 years. An additional protocol to the 1976 Rhine treaty was signed which limited the chloride concentration at the German–Dutch border to 200 mg/l.

The effectiveness of the international environmental protection of the Rhine can also be measured in terms other than the reduced number of chemical substances in the water, the increased amount of fauna and flora, or the extent to which international agreements were implemented. It can also be established by the support for the Rhine Action Programme proffered by the non-governmental actors in the regime. Both firms and environmental groups have endorsed and lauded the international efforts.

A last indication of the strength of the cooperation in the Rhine regime during the period 1987–95 is the extent to which the Rhine agreements went beyond national water laws and European directives. While before 1987 the Rhine agreements trailed behind national and European measures, after the Sandoz fire the Rhine plans led the way in water policy in the riparian states and in Brussels. It is of course difficult to establish, beyond a shred of doubt, the causal influence of the Rhine Action Programme on national and European legislation as many other factors were also at work, but three arguments may prove convincing. To start with, the Rhine Action Programme only stated goals and – through a series of reports from the ICPR – suggested methods. The way in which the goals were to be reached was explicitly left to national and local authorities. Therefore, it was the purpose of the RAP to stimulate national water policy. A further indication of the causal influence can be found in the timing of the RAP and national and European regulations. The concerns and themes of the RAP (especially its quality aims and its emphasis on ecosystem management) appeared a few years later in national and European legislation.[51] Last, during the interviews, government officials acknowledged the influence of the Rhine programmes on domestic policy and law. For instance, an official from the Unit Water Protection at Directorate-General XI of the EC stated that the Commission had come to regard the Rhine regime since 1987 as an example for water management all over Europe.[52]

When asked their views of the driving force behind the implementation of the RAP, a large majority of interviewees said that it was the friendly, trusting relationships they had built among each other that had made cooperation possible. This seems a crucial point. The actors themselves stated that before the Sandoz incident international cooperation was difficult to achieve as there was no trust between the involved organizations. Thereafter, or so the actors themselves asserted, the confidence that they had in each other made international cooperation on the protection of the Rhine possible.[53]

In sum, the infusion of the Rhine regime by Minister Kroes and consultants from McKinsey-Amsterdam with a 'let's get things done' attitude broke the deadlock that the international environmental protection of the Rhine had always been, and set off a truly cooperative and comprehensive international effort. Such a cooperative attitude is characteristic of the individualistic way of life.

After 1987, the governance of the international Rhine regime (the second regime principle/norm distinguished here) also became infused with individualism. According to the individualistic way of life, intergovernmental regulation should be kept to a minimum. Only those social problems that absolutely cannot be resolved without intergovernmental agreement should be regulated at the interstate level. All other issues should either be dealt with at the domestic governmental level, or (even better) should be left to the interactions among citizens themselves. This individualistic ideal was taken up in the international Rhine regime in 1987, after having been formulated and promoted by the McKinsey team of consultants. The RAP was a conscious attempt to lay the responsibility for the cleanup of the Rhine at the lowest possible governmental level: the German *Länder*, the Swiss cantons and the French *agences de l'eau*. The RAP was not a formal international treaty or convention, as the 1976 Chemical and Salt Treaties had been. Instead, it consisted of a report published by the ICPR, proposing a limited number of far-reaching environmental goals. The national delegations had only verbally endorsed these proposals. By keeping the Rhine Action Programme informal and non-binding, the national governments of the Rhine countries, as well as the European Commission, did not need to sign the document. Implementation was therefore voluntary and was left as much as possible to local levels of government. This form of minimal international coordination exemplifies the international governance preferred by individualism.

Another manifestation of the individualistic governance of the Rhine regime during 1987–95 was the decision to limit the size of the ICPR. In the course of its history, the structure of the ICPR had continuously expanded. It had come to include some 18 working groups, expert groups, subgroups and *ad hoc* groups, besides the plenary sessions and the meetings of the heads of the delegations. In 1994, the ministers agreed to downsize and simplify the structure of the ICPR. Since then, the ICPR has consisted of three permanent working groups, as well as two *ad hoc* groups. In an interview, a representative of the Dutch delegation to the ICPR referred to this action as 'adopting the principles of lean management' which was 'in keeping with the spirit of times'.[54]

Rules and procedures

As we have seen, the adoption of individualistic principles and norms regarding international cooperation and governance broke the deadlock that the environmental cooperation on the Rhine had become. This released the power of the ICPR to develop and propose a whole array of hierarchical environmental protection measures. The execution of the RAP by national and subnational authorities was supported by a barrage of proposals and suggestions for the environmental management of the Rhine catchment area. From 1987 until 1994, the ICPR published more than 60 research reports on the ecological revival of the Rhine. Although these ideas were merely proposals, they were nevertheless often taken up by the authorities all along the river. The proposals can be termed hierarchical, as they flowed from the assumption that ecosystems can be managed and mastered on the basis of expert knowledge. Below, the content of the environmental management system devised by the ICPR will be briefly described.[55]

The Rhine Action Programme aimed at a 50 per cent reduction of the emissions of 45 priority substances. To assist the governments in reaching this goal, the ICPR proposed the adoption by local authorities of a large number of water quality standards. In 1991, the ICPR calculated 59 strict water quality standards,[56] and local governments in the Rhine countries often used these standards as a basis for their policies. The RAP also introduced the theme of ecosystem management in the Rhine regime, concerning which the ICPR promoted a number of projects. Such higher species as salmon and sea trout could no longer live in the Rhine because their way up and down the river was blocked by dams and weirs, and because their spawning grounds had been destroyed by canalization and diking. The ICPR therefore advocated the construction of fish ladders and sluices at the dams and weirs. In addition, it proposed the redevelopment of a new stock of Rhine salmon and, at the behest of the ICPR, salmon eggs were purchased in Scotland and the southwest of France. The eggs were first hatched in special hatcheries and thousands of fry were released into the river. The ICPR developed a programme to monitor the behaviour of the salmon, and also studied and advocated the redevelopment of nature along sections of the riverbank. Such nature development projects would not only benefit migratory fish. Many animal species that once had lived alongside the river, such as the otter and various waterfowl species, had disappeared when canalization and the use of riverbanks for agriculture and urbanization had destroyed their natural habitat. The nature redevelopment favoured by the ICPR would also benefit these animals. Regional authorities within the Rhine countries have begun to implement some of these elaborate plans for ecosystem management.

In sum, between 1987 and 1995 a highly praised environmental regime for the Rhine existed that was based on individualistic principles and norms, complemented by hierarchical rules and programmes. Clearly a change *of* regime took place. Not many IR approaches could easily account for the rapid and fulsome change of the Rhine regime. Realism has often been criticized for its inability to explain profound change, and alternative frameworks have sought to explain the emergence of international cooperation in terms of the development of transnational relations, compliance mechanisms, administrative capacity, epistemic communities and scientific knowledge.[57] However, these explanations are not applicable to the transformation of the Rhine regime, as the aforementioned factors either did not exist or did not greatly change around 1987, the turning point in the history of the Rhine cooperation.

Here the fundamental change of the Rhine regime has been explained in terms of the shock that the Sandoz incident represented, the ensuing crisis in which the media, environmental groups and ministers became much more active, and the clever way in which the crisis was used by Minister Kroes and a team of McKinsey consultants to promote radically new forms of cooperation and governance in the Rhine regime. This is all in line with the writing on policy change by Peter Hall, Paul Sabatier and Oran Young, and grid-group theory enriches the story that can be construed on the basis of their writing. By spelling out the cultural orientations of policy brokers, it clarifies the *direction* of policy change. Ms Kroes and Dr Winsemius are both prominent members of the Dutch liberal party, the VVD. McKinsey consultants can also be expected to think in more individualistic terms than many public servants. As a consequence, Ms Kroes and the McKinsey team tried to reorganize the Rhine regime along individualistic lines.

1995 and beyond: towards an international regime based on three ways of life

Since the 1970s, egalitarian ideas about the protection of the Rhine basin have been expounded by a number of environmental organizations.[58] Regarding regime norms, these ideas have included a preference for more openness and public participation in international decision-making. Concerning rules, environmental organizations have favoured protection based on minimal human intervention with the ecosystems of the basin. For instance, although environmental organizations have applauded the adoption of the priority list of 45 substances from the Rhine Action Programme, they have also insisted that these priority substances include only about 5 per cent of all chemical substances dumped into the river by industry and agriculture. They claim that the rest of these substances may also be harmful to the environment and should be the subject of investigation and strict regulation. In their view, a complete rehabilitation of the Rhine basin would necessitate an overhaul of capitalist methods of production: eco-centrism instead of ego-centrism should be instituted.

One specific point of environmental groups has concerned the adverse consequences of the canalization of the Rhine. Canalization has cut off the river from its natural floodplains and has forced the water into a single, narrow stream. As a result, the water level of the river has risen and the river basin has become more prone to flooding. Environmental associations have also asserted that in losing its natural floodplains, the river has lost much of its capacity for self-purification. Water that flows back into the river from an alluvial terrain is purified by the ground of the floodplain which retains and dissolves pollutants. Since the beginning of the 1980s, environmental associations have tried to raise public awareness of the benefits of the reservation of lands for floodplains. Starting in 1984, for instance, the World Wide Fund for Nature – Germany (one of the least radical environmental groups in the area) has operated its own mini-alluvial plain to illustrate the concept.

Environmental associations have claimed that the massive inundations of the Rhine and Meuse (the river that flows from France, through Belgium, to the Netherlands) at the end of 1993 and the beginning of 1995 vindicated their demands for the restoration of natural floodplains. The floods led to the evacuation of a number of cities and towns while the total cost of the inundations has been estimated at several billion German marks.[59] The overflows came as a huge surprise to most people, and the immediate damage in financial terms was much larger than that of the Sandoz accident. As the floods came as such a surprise and can be construed as justifying the egalitarian discourse of environmental groups, it seems reasonable to expect that they triggered a transformation of the Rhine regime along egalitarian lines.

Initially, such a change of the Rhine regime did not take place, for which various causes can be outlined. First, the floods were not widely blamed on existing government policies. In the media, the overflows were presented as having been caused by both extremely bad weather conditions and river banks that had been too low. To avoid future disasters, it was argued that the existing dikes along the river should be fortified. These opinions only strengthened the hierarchical project of mastering and subjugating the river: the media put much less emphasis on the alternative argument offered by environmental groups, namely that the dikes had not been too low in some places, but instead had been too high in others.[60] The inhabitants of the Rhine area seemed to agree with the opinions expressed in the press – no mass demonstrations were staged to denounce government policies as had been the case after the Sandoz fire. On the contrary, the only backlash that seems to have occurred was directed against voluntary associations. In the Netherlands, the houses of several environmental activists had to be put under police surveillance after death threats had been issued by disgruntled house owners.

At the governmental level it was also not felt that the floods had invalidated existing water policies. On the one hand, it was acknowledged that dikes had to be heightened in various parts of the Rhine basin, for which plans were swiftly drawn up and implemented. On the other hand, the governments of the Rhine countries recognized the need to restore some of the floodplains of the river.[61] However, they also felt that this would require a mere continuation of existing policies, especially of the plans to redevelop ecosystems along the river banks. As the 1993 and 1995 floodings of the Rhine and the Meuse were not widely blamed on existing government policies and underlying rationales, but instead strengthened the support for these policies, they did not immediately lead to a transformation *of* the regime to protect the Rhine. No principles or norms of the regime were directly affected. The floods merely led to a change *within* the regime, in the form of more efforts to restore alluvial plains as well as new plans to strengthen and heighten dikes.

At the Ministerial Conference of 1994 the Dutch Minister, Ms Jorritsma, advocated for the inclusion of non-governmental organizations (including both firms and citizens' groups) in the Rhine regime. In particular, she proposed to give NGOs observer status. In defending her proposal, she did not refer to the fact that environmental organizations had foreseen the floods – the Dutch delegation to the conference felt that such a strategy would have been 'inappropriate'. In their eyes, it is against the mores of intergovernmental negotiations to confront other national delegations with errors of judgement that they may have made.[62] Instead, the Dutch side used two other arguments. First, it argued that inclusion of NGOs might increase public acceptance of, and support for, international measures. Second, the Dutch also pointed out that NGOs had been welcome at the international conferences on the environmental protection of the North Sea since 1992. The Dutch proposal was opposed by all other governments. It was pointed out that, in contrast to the North Sea conferences, the Rhine negotiations merely involved five countries plus the European Union. If NGOs were allowed to participate in official meetings of the ICPR, their representatives might outnumber governmental officials. It was also pointed out that there was no need to win any public support, since most NGOs had enthusiastically endorsed the Rhine Action Programme. Most importantly, the delegations from France, Germany and Switzerland felt that it was the prerogative of governments to take international decisions. However, in the atmosphere of cooperation and amiability that had blossomed since 1987, it was also felt that a concession should be made to the Dutch government, which entailed that the ICPR would discuss new policy plans with representatives from environmental associations, industry, and agriculture. It would also inform these organizations of any decisions that had been taken. In addition, NGOs were offered the opportunity to present their opinions to the yearly plenary meetings of the ICPR.

So, in the short run, the Rhine floods did not lead to a significant transformation of the Rhine regime. Cultural theory is useful in understanding why not. By setting out ways of life, it is able to shed some light on which shocks contradict the rationalities followed by dominant actors, and which shocks these rationalities can accommodate. The 1993 and 1995 floods did not immediately lead to an overhaul of the Rhine regime, partly because it was felt that ongoing principles and policies would be sufficient to prevent future inundations. Other approaches, which do not a priori spell out ways of life/rationalities, run the risk of offering circular explanations. Such approaches would designate a shock that did not lead to fundamental policy change as not 'shocking enough'.

Yet, only a few years later new principles and norms were added to the Rhine regime. In 1997 it was decided not only to give NGOs observer status at the Ministerial Rhine Conferences, but even to delegate implementation of some international policies to NGOs. These decisions were taken up in the new Convention for the Protection of the Rhine that was signed by the responsible ministers in January 1998. These novel principles and norms are clearly egalitarian. It is an egalitarian ideal to take international governance out of the hands of impersonal bureaucracies and to let transnational issues be decided by those who are directly affected. This introduction of egalitarian principles represents another change *of* the Rhine regime. It meant an opening towards an international regime based on three ways of life: hierarchy, individualism and egalitarianism. Again the swiftness of the transformation was conspicuous. Before 1996 the ICPR had had close to zero contacts with NGOs. In 1998, NGOs were asked to carry out certain tasks of the ICPR. What can account for this rapid change, if it was not the surprise of the floods?

A combination of factors was decisive,[63] several of which were external to the Rhine regime. Starting in the mid-1980s, the domestic authorities in all Rhine countries had begun to relax their stringent command-and-control approaches to environmental protection, and part of this development had involved more consultation with NGOs, including consultation on the matter of water protection.[64] Moreover, other international commissions dealing with transboundary pollution had already granted observer status to NGOs. Thus, with regard to the issue of public participation, the closed ICPR had increasingly become the odd one out, and in the light of these domestic and international trends, it had become more and more difficult to refuse observer status to NGOs at the ICPR meetings. The German delegation to the ICPR was especially swayed by these arguments and started to endorse the position of the Dutch government in 1997. External factors like these have not been pointed out in cultural theory, although they have been highlighted by Paul Sabatier.

An additional factor was internal to the Rhine regime. In March 1996 the ICPR had its first official meetings with NGOs – a result of the concession made to the Dutch minister at the Rhine Conference of 1994. As all NGOs (except for agricultural representatives) praised the revival of the Rhine, this meeting was considered to be a huge success by the governmental delegations which eased the qualms that the French and Swiss governments had had about granting observer status to NGOs. In 1997 it therefore became possible to persuade the French and Swiss governments to allow NGOs (water supply companies, farmers' associations, industrialists and environmental groups) to sit in at the Ministerial Rhine Conferences. The first such 'open' conference was held in January 1998. Again, the deliberations with NGOs were valued quite positively. The governments discovered that NGOs were not only supportive of the Rhine policies, but that they were also able to provide constructive suggestions for ecosystem management. This further bolstered the cooperation between the ICPR and the involved NGOs.

This last, internal factor can be described in the language of cultural theory. It can be understood as a discovery of common viewpoints by organizations following alternative ways of life. Hierarchical rules and procedures for river management come in two forms. One of these forms to a large degree overlaps with the rules and procedures favoured by egalitarian organizations, while the other is heresy to believers in egalitarianism. The latter set of hierarchical rules and procedures consists of plans for mastering and exploiting rivers. These plans include canalization, as well as the construction of dams, dikes and weirs. These programmes are hierarchical as they rest on the assumption that humans can control and tame nature on the basis of expert knowledge. This has been the traditional way in which governments have tried to manage river basins,[65] but it has come under ferocious attacks from more egalitarian-minded environmental activists who have argued that it is neither feasible nor just for humans to exploit natural systems in this way.[66] It has been appreciated by individualists for providing the infrastructure that has allowed them to truck, barter and exchange. The second set of hierarchical rules and procedures for the management of water basins is more recent. These are the rules and procedures that make up ecosystem management – a holistic approach to environmental protection that emphasizes the mutual relations between all the elements of a natural system. Ecosystem management provides ample opportunities for hierarchical policies.[67] It calls for endless studies of how the flora and fauna of ecosystems interrelate, it necessitates sophisticated measurement systems, and it involves the development of complex programmes for restoring the precarious balances between animals and plants. These measures can be called hierarchical, as they privilege a select group of experts – those who are knowledgeable about ecosystems. Yet, ecosystem management has also been a favourite among egalitarians. This is because it is based on human respect for natural systems, which is a far cry from hierarchical attempts to master and exploit nature. It puts animals and plants more on a par with humans, something coveted by egalitarians. So ecosystem management can be seen as a 'focal point' around which egalitarian and hierarchical organizations have coalesced in environmental politics. It could be argued that this new coalition has started to replace the erstwhile project of individualistic and hierarchical organizations to master nature. That is to say, in environmental politics hierarchical and egalitarian organizations have started to agree on rules and procedures, although they still favour alternative principles and norms. Yet, these agreements concerning rules and procedures of environmental management have at the same time made it more likely that more hierarchical governments are willing to accept and adopt some of the principles and norms favoured by more egalitarian-minded environmental organizations.

This process has also taken place in the Rhine regime. Through implementation of the Rhine Action Programme, the ICPR has taken its first steps towards a full-fledged ecosystem approach. It has endeavoured to restore alluvial plains along the Rhine, and to increase fish stocks in the river. The new Convention for the Protection of the Rhine (that was signed in January 1998) strengthens these measures. To quote the ICPR, the new Convention fixes the following targets for international cooperation along the Rhine: sustainable development of the Rhine; further improvement of the ecological state; holistic flood protection and defence taking into account ecological requirements; and the preservation, improvement and restoration of natural habitats and of the natural stream function.[68] These initiatives contain many measures that the more egalitarian-minded environmental organizations have been asking for since the 1980s. It is therefore not surprising that environmental NGOs have been especially supportive of the international policies for the protection of the Rhine that have been developed by the ICPR in recent years. The mutual agreement between environmental groups and authorities on the rules and procedures that should make up the Rhine regime has also facilitated the introduction of more egalitarian principles and norms. It induced the relevant French and Swiss ministers to accept Dutch and German proposals to let NGOs play a part in deciding upon, and implementing, international Rhine policies. The 1993 and 1995 floods have only played an indirect role in this regime change. They have speeded up as well as expanded government plans to restore alluvial plains along the Rhine – a central element in the ecosystem management of the river for which environmental groups had been asking for a long time. In this manner, the floods of the Rhine and Meuse have brought the policy views of NGOs and ministries closer together, which also set off changes in the principles and norms of the Rhine regime. In 1998 an international regime that mixes elements of three cultures (hierarchy, individualism and egalitarianism) therefore came into being. In principle, such an arrangement comes close to cultural theory's normative ideal for decision-making, although at the moment it is much too early to assess the consequences of the NGO access to the ICPR.

In summary, the massive floods of the Rhine and Meuse during 1993 and 1995 did not lead (at least not directly) to a radical transformation of the Rhine regime along egalitarian lines. The initial reaction of governments was to keep NGOs out of international negotiations, despite the fact that the inundations came as a huge surprise to most of the people and government officials affected, and despite the efforts of the Dutch Minister for Transport, Public Works and Water Management to include NGOs into the structures of the ICPR. In other words, despite the occurrence of both shocks and policy entrepreneurs, no overhaul of the Rhine regime took place. Yet, surprises and policy brokers are the two conditions for radical policy change that are emphasized by Hall and Sabatier. Cultural theory sheds some more light on why the Rhine regime was not fundamentally altered. It spells out which principles, norms, rules and procedures actors prefer. In doing so, it can clarify which shocks can be accommodated by dominant rationales and which cannot. After the floods of the Rhine and the Meuse in 1993 and 1995 it was widely felt that future floods could be prevented with ongoing environmental policies and principles, and, if anything, the 1993 and 1995 floods discredited alternative policy views. These floods, therefore, did not lead to a radical change of the Rhine regime. Just a few years later, however, NGOs were allowed to play a much more active role in the protection of the Rhine which constituted a veritable change of the Rhine regime. This change was partly caused by several factors external to the Rhine regime, but the transformation was also partly due to the fact that the more hierarchical government organizations and the more egalitarian environmental movements had discovered common ground with the development of the ecosystem approach to the Rhine basin.

Conclusion

Many authors have claimed that fundamental policy reversals should be explained on the basis of shocking events that could neither have been foreseen nor solved with the rationales followed by the dominant actors in the issue-area. Yet very few authors have attempted to spell out which rationales actors adhere to. Grid-group analysis is an exception. The approach does spell out which rationales actors tend to believe in and, therefore, grid-group theory is able to shed more light on why some surprising events have led to fundamental policy change, while other shocks have not triggered such a transformation of an issue-area. Furthermore, it can provide a fuller answer to the question: why did the issue-area change in the way it did (and not in other conceivable ways)? In this chapter these ideas have been exemplified with the history of the international regime to protect the Rhine river. In doing so, it was felt necessary to combine cultural theory's conceptualization of change with several ideas expressed by Peter Hall, Paul Sabatier and Oran Young.

This chapter has also illustrated the thesis that the ways of life distinguished in cultural theory matter. People and organizations who have adhered to alternative cultures have perceived both the problems of, and the solutions to, the degradation of the Rhine in quite different ways. In particular, those who have tended towards individualism have been much less disturbed by the existence of state borders than those who have tended to think and act along hierarchical lines. The former have advocated a pragmatic, informal and trusting approach to international cooperation, whereas the latter have insisted on legalistic, top-down and slow-moving international procedures. This does not of course mean that international environmental cooperation should always and ever be left to those believing in the individualist way. If that were so, one problem, for instance, would be that individualists tend to deny or ignore accumulating evidence of environmental degradation. In the case of the Rhine, the Sandoz incident opened the eyes of the responsible Dutch Minister, Ms Neelie Kroes, a member of the most individualistic major political party in the Netherlands.

The chapter has also exemplified the proposition that 'more ways of life may be better'. The 'monocultural' Rhine regime that existed before 1987 was clearly inferior to the 'bicultural' regime that existed between 1987 and 1995, and the 'tricultural' regime that came into being thereafter.

5
Who has Washed the River Rhine?

Introduction

In 1834 Samuel Coleridge asked:

> In Köln, a town of monks and bones,
> And pavements fang'd with murderous stones
> And rags, and hags, and hideous wenches;
> I counted two and seventy stenches,
> All well defined, and several stinks!
> Ye Nymphs that reign o'er sewers and sinks,
> The river Rhine, it is well known,
> Doth wash your city of Cologne;
> But tell me, Nymphs, what power divine
> Shall henceforth wash the river Rhine?[1]

Though not the product of nymphs, this research has allowed an answer to Coleridge's question. Swiss, French, German and Dutch industries along the Rhine have made huge and voluntary contributions to the clean-up of the Rhine. Companies located in the Rhine valley have made costly investments in water protection (including the building of sewage treatment systems) that exceed national water legislation and go far beyond international measures. Another significant force for the restoration of the Rhine has comprised the domestic authorities in the riparian countries. In particular, domestic water policies developed at both the national and the local level have enabled cities along the Rhine to build sewage treatment plants. Domestic governments have also cooperated with the chemical industry in its efforts to clean up discharges into the river.

The extensive attempts by the governments of the riparian countries to conclude and implement international agreements on the ecological restoration of the river have been much less effective. Until the Sandoz incident in late 1986, intergovernmental efforts were at best ineffective and at worst positively harmful for the ecological revival of the Rhine. Only in 1987, after the adoption of the Rhine Action Programme by the governments involved in reaction to the Sandoz incident, did international measures begin to exert a positive influence on the ecological conditions of the Rhine basin. But even before the adoption of the Rhine Action Programme, firms and domestic authorities had cleaned up the Rhine to a large degree. The agricultural sector has been an even more uncooperative actor. While industries and communities along the waterway have drastically reduced their emissions of toxic substances since the early 1970s, farmers in the Rhine catchment area have continued their discharges of toxic matters into the water. This has taken place despite increasingly stringent domestic and European legislation aimed at reducing emissions from diffuse sources.

This chapter will attempt to explain why the various public and private organizations that are involved in the protection of the Rhine have made widely divergent contributions to the clean-up of the river. First, this puzzle will be presented in greater detail – describing (among other things) the extent to which chemical firms and communities along the river have taken protection measures, as well as the degree to which the agricultural sector has refused to contribute to the clean-up of the Rhine. Subsequently, various explanations will be considered that could be offered for these developments. The approaches that will be discussed include neorealism, neoliberalism, the 'common pool resources' model developed by Elinor Ostrom and colleagues, as well as a specific analysis of the protection of the Rhine undertaken by Thomas Bernauer and Peter Moser. It will be argued that none of these approaches can adequately explain the differences in cooperativeness between the actors concerned. Thereafter, an alternative explanation will be offered that is based on the ways of life followed by the various groups of public and private actors. This explanation will provide another example of a remarkable transboundary process that also played an important role in the events described in the previous chapter. This transboundary process can be formulated as follows. Organizations that tend to follow the individualist way of life are less hampered by the existence of state boundaries than organizations that adhere more to the hierarchical way of life. As such, the former organizations are sometimes inclined to make more extensive contributions to the solution of transboundary problems than the latter. In Chapter 4, the acceptance of individualist principles and norms that had been promoted by a team of McKinsey consultants broke the deadlock of intergovernmental negotiations on the protection of the Rhine. The present chapter will show that the individualist way of thinking and acting in which Rhine companies have tended to engage to a large degree explains why these organizations have made more extensive contributions to the restoration of the river than other organizations. The account of the environmental restoration of the Rhine that will be offered below also clearly demonstrates that an international regime does not solely (or even predominantly) consist of intergovernmental relations. The principles, norms, rules and procedures that are followed by private actors and domestic authorities play a huge role as well.

The puzzle

Some years ago, reflecting upon the possibilities for environmental restoration in the present world system, Oran Young stated:

> Under a variety of conditions, the operations of governments are not only insufficient to ensure that growing demands for governance are met but also may be unnecessary for the provision of governance. This realization is truly liberating.[2]

The possibilities (and limits) for such 'governance without government' are well-illustrated by the case of the environmental protection of the Rhine. At present, the Rhine is widely regarded as the cleanest of all the main waterways in Europe.[3] Scientists, representatives from environmental groups and firms as well as officials working for national governments and the European Union are in agreement on this. The ecological conditions of the Rhine have been systematically monitored by several public and private organizations: the International Commission for the Protection of the Rhine (ICPR), the Deutsche Kommission zur Reinhaltung des Rheins (DKRR), the Internationale Arbeitsgemeinschaft der Wasserwerke in Rheineinzugsgebiet (IAWR, the umbrella organization of the water supply companies along the Rhine) and the chemical industry (Henkel, BASF).[4] These records show the trends in the levels of chemical substances in the Rhine and the development of parameters for overall chemical pollution as well as changes in biological parameters (such as biodiversity and oxygen levels). The records indicate that the restoration of the Rhine has taken place over a period of some 25 years. In the late 1960s and early 1970s, the ecosystems of the river were in an extremely bad shape as a consequence of two processes. The first concerned the discharge of toxic substances into the water by firms, farms and municipalities. The second consisted of efforts to increase the economic usefulness of the Rhine, including canalization of the river as well as the building of dams and weirs. Since the end of the 1960s, massive investments in sewage treatment plants by industry and cities have greatly improved water quality. The greatest strides were made in the 1970s and early 1980s. Even by 1986 the discharge of many suspected chemical substances had been reduced by about 80–90 per cent. In subsequent years, even more reductions were undertaken. After 1986, a start was made with a full-fledged ecosystem approach.[5] Under this approach, concern is given not only to water quality, but also to the links between water quality and water quantity, as well as to the relationships between flora, fauna and water quality. Concrete measures have been the construction of fish ladders around dams and weirs (to give free passage to migratory fish), the reintroduction of salmon and the reconstruction of alluvial plains and spawning grounds along the river.

Different groups of actors have made widely divergent contributions to the environmental restoration of the Rhine basin. In brief, the large industrial companies have made the most extensive efforts to clean up the Rhine, closely followed by domestic authorities and municipalities. Much less helpful have been the governmental delegations to the ICPR and, especially, the agricultural sector. These differences in cooperative behaviour form the puzzle of this chapter: *what can explain the vast differences in cooperativeness with regard to the restoration of the Rhine between the relevant groups of public and private actors?* A comparison between the environmental protection efforts of governmental and non-governmental organizations is usually difficult to make, as these organizations have different objectives. In this particular case it is possible to make such a comparison, as the environmental protection measures of the large industrial companies in the Rhine valley have gone far beyond existing domestic and international legislation, while the protection measures taken by farms have lagged behind existing legislation. This makes it possible to compare the activities of firms and farms with the efforts of governments. Below is set out in detail the contributions that the various groups of organizations have made to the environmental protection of the Rhine.

Large industrial firms

Many industrial firms are located in the Rhine valley. The Ruhr area is famous for its steel production and mining, while about 20 per cent of the world's chemical industry is situated along the banks of the Rhine. The chemical firms in the Rhine catchment area include both giants such as Hoechst (Frankfurt), Bayer (Leverkussen), BASF (Ludwigshaven), Shell (Rotterdam), Ciba-Geigy and Sandoz (Basel),[6] and a large number of smaller companies. Smaller enterprises release their effluents into the sewage treatment plants of the communities in which they are located while large firms have had to build their own wastewater facilities. These large industrial corporations have made huge contributions to the clean-up of the Rhine. As mentioned in Chapter 4, in all riparian countries firms have been obliged to acquire wastewater discharge permits from governmental departments which have stipulated the minimum criteria for the effluents of each specific company. These criteria have been based on industry-wide effluent limits, as well as water quality standards developed by the ministries responsible.

From the mid-1960s onwards, the large companies along the Rhine have invested in water protection. In an extensive attempt to purify their effluents, they have built sewage treatment plants and have made changes in their production processes.[7] The first generation of industrial sewage treatment plants came into operation around 1975 and some ten years later these plants were replaced by a newer and more sophisticated generation of treatment plant. In the early 1990s, a third generation of sewage facilities was built by the large companies in the Rhine basin. These rounds of investments in water protection have been very costly. For instance, it has been estimated that the German chemical industry in the Rhine basin spent DM 6.6 billion on building and operating sewage treatment plants between 1987 and 1991 alone.[8] The records of the ICPR and the IAWR show that in the period between 1970 and 1986 the emissions of many suspected chemical matters by industries were reduced by about 80–90 per cent, while as early as 1979 an independent research team from the University of Stuttgart concluded that 'the variety and size of the fish stocks at the site of BASF are normal. The fish are extraordinarily healthy.'[9]

It is one thing to say that industrial firms have taken extensive environmental measures. It is quite something else to argue that these measures went beyond governmental policies. To make the latter statement plausible, a comparison between the efforts of companies and the goals of governmental policies is needed. Ideally, such a comparison would confront the toxicity of the effluents of Rhine companies with the effluent limits imposed on firms by the responsible ministries. Unfortunately, it is not feasible to undertake this direct comparison as no public data exist on the discharges of individual Rhine firms. However, a slightly more indirect comparison is possible. The water quality standards that have been formulated by the involved ministries of each riparian country can be set off against the water quality data assembled by the IAWR and the ICPR. The problem is that these water quality standards have varied from one Rhine country to another. Only in 1991, as part of the Rhine Action Programme, did the governments of the riparian countries adopt a common list of water quality standards, but these common standards can be compared to the actual quality of the Rhine water in the mid-1980s. This test of the extent to which the Rhine companies have made voluntary investments in water protection is quite a stringent one, as the standards adopted in 1991 went much further than the domestic water quality standards existing at the time in the riparian countries. Table 5.1 shows the results of this comparison.

Table 5.1 shows the amazing extent to which industrial firms along the Rhine have made voluntary investments in water protection. In 1991 the ICPR adopted 59 water quality standards for the whole Rhine basin that were stricter than the water quality standards existing within the riparian countries. Several of these standards concern chemical substances that are only released by agricultural firms which are not of relevance here. However, for 22 of the remaining substances, the ICPR and IAWR have assembled data. Table 5.1 makes clear that *even by 1986* the levels of all but four of these 22 priority substances in the Rhine water were lower than the strict water quality standards adopted by the governments *in 1991*. This means that the discharges into the Rhine had already been greatly and *voluntarily* reduced by the mid-1980s. This result is all the more impressive when one realizes that the IAWR data used in the table refer to the year 1986. This was the year in which the Sandoz incident, as well as many other accidental spills of chemicals, took place.[10] In addition, the result cannot be denied by arguing that these international

Table 5.1 A comparison between water quality standards for the Rhine and the actual water quality of that river at Lobith (both in µg/l, unless otherwise indicated) (the lower value is expressed in bold)

	Water quality standards of 1991	Actual water quality in the mid-1980s
Mercury	0.5 mg/l	0.00005 mg/l
Cadmium	1.0 mg/l	0.0001 mg/l
Chromium	100.0 mg/l	0.008 mg/l
Copper	50.0 mg/l	0.006 mg/l
Nickel	50.0 mg/l	0.005 mg/l
Zinc	200.0 mg/l	0.048 mg/l
Lead	100.0 mg/l	0.005 mg/l
Arsenic	40.0 mg/l	0.002 mg/l
DDT (various compositions)	0.001	<0.001*
Endosulfan	0.001	<0.001*
α-HCH	0.1	<0.01
γ-HCH	**0.002**	0.02
Pentachlorophenol	0.1	**0.03**
1.2-Dichloroethane	**1.0**	2.3*
Trichloroethene	1.0	**0.1**
Trichloromethane	0.6	**0.2**
Tetrachloromethane	1.0	**0.2**
Chloronitrobenzene	1.0	**[between 0.04 and 0.09*]**
Trichlorobenzene	0.1	**0.03***
Hexachlorobenzene	0.001	<0.01
PCBs	**0.0001**	0.006
Ammonium-N	**200**	670

Sources: ICPR, *Statusbericht Rhein* (1993), pp. 118–20 (Rhine water quality standards); RIWA (Samenwerkende Rijn- en Maasbedrijven, the Dutch member-organization of the IAWR), *Jaarverslag 1986 – deel A: de Rijn* (1986), pp. 80–5; ICPR, *Year Report (Tätigkeitsbericht)* (1985), pp. 196–7. An asterix (*) indicates a value taken from the ICPR year report. The ICPR figures refer to the year 1985. The other numbers (taken from the IAWR/RIWA publication) concern the year 1986.

water quality standards were less stringent than domestic ones. In fact, these ICPR standards were stricter than the domestic water quality standards at the time (1991), let alone in the mid-1980s. Table 5.1 leaves room for only one interpretation: the impressive reduction of the pollution of the Rhine has been mainly the result of voluntary measures in water protection by point source dischargers.

It cannot even be said that it has been the aim of the authorities within the Rhine countries to induce point source dischargers to take voluntary measures. During the interviews, only one government official[11] (out of a total of 34 government employees) acknowledged the voluntary water policies developed by point source dischargers. A large majority of government officials stated that it was their belief that firms would only invest in water protection when forced to do so by legal means. The adoption of water quality standards and emission norms was firmly based upon a belief in the necessity of a strict command-and-control approach to pollution prevention.

A last counterargument also does not hold. Firms and communities within the Rhine area have had to pay effluent fees on their wastewater discharges into the river so it could be argued that these fees have prompted firms and municipalities to make investments in water protection which could then not properly be seen as 'voluntary'. The argument would seem especially relevant for companies, as cities in the Rhine countries have not only paid effluent fees to the responsible authorities, but have also received a lot of financial aid from the governments to construct wastewater facilities. It is therefore especially interesting to consider the 'effluent fees mattered' thesis with regard to companies. The fact of the matter is that the effluent fees paid by enterprises in the Rhine watershed have always remained small. A few figures illustrate this. In the Netherlands, industry paid 348 million guilders in effluent fees in 1985 and 611 million guilders in 1996. This came to 0.3 per cent and 0.4 per cent of the production costs of industry in these years. In Germany, the amounts involved have been even smaller. For instance, in 1995 the firms in *Länder* through which the Rhine flows paid approximately DM 86.1 million in effluent fees.[12] In terms of the added value of these companies this comes to about 0.014 per cent. The number also pales into insignificance compared to the DM 6.6 *billion* incurred for water protection by just the German *chemical* companies between 1987 and 1991.[13] Moreover, picking such a late year (1995) does not underestimate the importance of effluent fees. Granted, discharges into the Rhine have been greatly diminished over time, thus reducing the amounts to be paid in effluent fees. However, in order to make up for declining revenues, the German authorities have raised effluent fees from DM 12 per pollution unit to DM 70 – an almost sixfold increase. It can therefore be concluded that German effluent fees have always been tiny, and therefore of very little consequence to German firms on the Rhine. In Switzerland, no effluent fees have been raised between 1970 and 1998. So, as the effluent fees in the Rhine basin have always remained small, they can only have played a correspondingly limited role in the decision-making of companies. As

Peter Rogers observes: 'In general, the [European] governments are reluctant to raise the effluent charges high enough to achieve the desired levels of waste reduction.'[14] The European pollution fees have been much too small to serve as a powerful explanation for why the Rhine companies have invested massively in water protection. Hence, these huge investments should still be regarded as being voluntary efforts.

Does this result not negate the findings of Chapter 4? In that chapter it was argued that the Rhine Action Programme (RAP) of 1987 was a success. The 59 water quality standards adopted by the riparian countries in 1991 were part of the RAP, but the above discussion shows that these standards had already been met in 1986, one year *before* the adoption of the RAP. Thus it could be said that this undermines the central message of Chapter 4 in two ways. First, it could be argued that the RAP was not so much of a success after all (since its pollution prevention aims had already been met from the outset). Second, it could be speculated that the adoption of the Rhine Action Programme did not have a lot to do with the activities of Minister Kroes and the McKinsey consultants, but had simply become possible because the actual quality of the Rhine water had already surpassed the standards that were included in the RAP.

It is possible to refute both these related objections. Pollution prevention was only one goal of the RAP; ecosystem management was another. In Chapter 4, it was shown that the RAP enabled the ICPR to make important steps towards realizing ecosystem management of the Rhine basin. While this was already a major achievement, with regard to pollution prevention the RAP probably also had an impact. Another pollution control aim taken up in the RAP (besides meeting the water quality standards) was a 50 per cent reduction of the 1985 emission levels of priority substances. This aim had by and large been reached by 1995.[15] Of course it could be argued that the point source dischargers would have further reduced their emissions anyway, also without the RAP. However, it is contended here that the Rhine Action Programme was at least one factor in the decision of point source dischargers to invest in a third generation of sewage treatment plants. Therefore, it can still be maintained that the RAP was a successful international initiative. This also invalidates the argument that the activities of Minister Kroes and the McKinsey consultants were not necessary for the adoption of the RAP.

Table 5.1 illustrates the extent to which point source dischargers voluntarily reduced their pollution. Point source dischargers come in two forms: large industrial firms and municipalities. The findings presented above indicate that both firms and cities have made extensive voluntary investments in water protection, and the efforts of both corporations and communities need to be acknowledged. Still, it is possible to argue that the large industrial firms in the Rhine basin have shown an even greater degree of cooperativeness than the cities in the Rhine area. First, several large chemical firms have developed revolutionary technology for the treatment of wastewater. In particular, the 'tower biology' developed by Bayer and the 'Bio-Hochreaktor' of Hoechst made reductions in water pollution possible that had hitherto been unthinkable.[16] Indeed, Bayer has sold its technology for sewage treatment worldwide. It should be borne in mind that investing in the development of these technologies has been risky and costly. The large chemical firms in the Rhine catchment area often set themselves environmental goals, and then proceeded to design the technology needed to reach these aims. Whether such technology could be developed was not at all clear when the environmental goals were selected by the firms.[17]

A second argument for the greater cooperativeness of firms (as compared to municipalities) is based on expert opinion. In 1992, the Internationale Arbeitsgemeinschaft der Wasserwerke in Rheineinzugsgebiet (IAWR, the umbrella organization of water supply companies) awarded its 'WRK-Rheinpreis'[18] to the German chemical firm Bayer AG in recognition of its exemplary contributions to the restoration of the Rhine.

A third argument is a financial one. The cities in the Rhine basin have received a lot of financial support for the construction of sewage treatment plants. Corporations in the basin have not received such support. In fact, firms have partly paid for the construction of sewage treatment plants in cities: their effluent fees have often been used by local governments to partly finance the construction of sewage treatment plants within cities. In this way, payments by Rhine firms have contributed to the reduction of pollution by communities. In sum it can be argued that, although the cities in the Rhine basin have made huge contributions to the restoration of the river, the firms located on the banks of the Rhine have made even bigger contributions.

It is also important to realize that Rhine companies have not attempted to switch pollution from one medium to another. This could have been attempted: in theory, it is possible to burn the substances that are kept out of the wastewater, thereby putting toxic chemical matter into the atmosphere. At the same time that large Rhine firms were developing water protection policies, they also heavily invested in similar air protection programmes.

National and local governments

The cooperativeness of the national and local governments with regard to the environmental protection of the Rhine has been Janus-faced. At the domestic level, increasingly comprehensive water protection policies have been developed in all Rhine countries since the late 1960s. Meanwhile, at the intergovernmental level very little progress was made until 1987, as bitter battles were fought between the national delegations to the International Commission for the Protection of the Rhine. The previous chapter has detailed both these processes, but Chapter 4 did not assess in full the effects of domestic and international policies on the ecological conditions in the Rhine catchment area. This will be attempted below, by looking first at the domestic efforts by national and local governments, and then at the international developments.

The governments of the Rhine countries began developing and implementing national water protection policies and laws at the end of the 1960s.[19] The basics of these domestic programmes have been quite similar. In each country, point source dischargers have had to acquire a permit for their emissions of effluents and the conditions attached to the permits have been calculated on the basis of water quality standards and effluent limits. In addition, point source dischargers have had to pay a pollution fee to the relevant government. Implementation of these water protection measures has usually been left to local government.

The water quality standards and effluent limits adopted in the riparian countries can only have had an indirect impact on the protection of the Rhine, since the point dischargers in the Rhine basin have usually set more stringent environmental targets for themselves. This indirect impact has consisted of the need to build sewage treatment systems in industrial plants and communities since pollution permits have only been given to point source dischargers who had acquired facilities for the treatment of wastewater. This has been an impetus for firms and cities to acquire such facilities. Municipalities seem to have been especially affected by this as city officials declared during the interviews that the construction of sewage treatment plants within their municipalities was a direct consequence of domestic water protection policies.[20] Firms, once again, seem to have taken a greater initiative on their own.

By highlighting the importance of water protection through legislation, by making it obligatory for firms and cities to build sewage treatment plants and by consulting with industries and towns, domestic policy-makers in the Rhine countries have made a significant contribution to the clean-up of the Rhine.

The intergovernmental efforts to protect the Rhine have been much less effective than domestic policies and laws. The same ministries that have been responsible for developing timely water protection policies within their own borders could not come to any international agreement concerning the restoration of the Rhine before the Sandoz incident in late 1986. As a consequence, the attempts to devise international agreements for the protection of the Rhine had no significant influence on the ecosystems of the Rhine basin before 1987. This was not for lack of trying. The argument could be made that the governments of the riparian countries could not agree on international measures, as domestic programmes were already sufficient. This argument does not hold. If it had been the case that the governments were simply not interested in international agreements, then why did they spend so much time and effort in devising such agreements? Also, in that case, how could so much acrimony have resulted from the lack of progress at the international level? (Recall that in 1979 the Dutch ambassador in Paris was recalled over the issue of salt discharges into the Rhine.) Moreover, why did the governments resort to international agreements after the Sandoz incident? These three counterarguments show that the governments of the Rhine countries were genuinely interested in protecting the river through international cooperation, but simply could not find enough common ground before the Sandoz spillage.

The Rhine Action Programme, adopted in 1987, has had a positive influence on the restoration of the river basin, but not nearly as much as is often claimed by government officials or representatives of environmental groups. As argued above, the RAP has led to a 70–90 per cent reduction of the 1985 emission levels of almost all priority substances into the river. However, in the years prior to the acceptance of the Rhine Action Programme much larger reductions had been achieved. Indeed, the bulk of improvements had taken place in the years 1970–86 when the emissions of chemical matter and heavy metals were reduced by 80–90 per cent. Imple-

mentation of the Rhine Action Programme has only been responsible for cutting back the remaining 10–20 per cent, although another
effect of the RAP has been the installation of more extensive systems to prevent accidental spillages.[21] However, it is not clear whether
the Rhine firms would not have taken these measures anyway after
the Sandoz incident without the RAP. In fact, the most important
influence of the RAP on the revival of the Rhine has been the introduction of ecosystem management. Concrete measures have
included the restoration of feeding grounds for migratory fish in
several places, the building of fish ladders at weirs and dams in the
river, and the redevelopment of the habitat of certain birds and
mammals (such as the otter) on the watersides.

Farms

Even more uncooperative than the international negotiators has
been the agricultural sector. The records of the IAWR show that
this sector has contributed very little to the clean-up of the Rhine,
a fact illustrated by the amount of nitrates in the river.[22] Nitrates
are components of the fertilizers used on farms and are chemicals
seen as highly toxic by the ICPR, the EC and national governments.
The governments of the riparian countries have attempted to reduce
the level of nitrates in open waters and groundwater in various
ways, including subsidy programmes.[23] The EC adopted a Nitrates
Directive in 1991,[24] and one of the main concerns of the Rhine
Treaty that is presently in preparation will be the reduction of the
discharge of nitrates and other chemicals used on farms into the
river. Despite these governmental efforts to reduce the use of
nitrates, the level of these chemical substances in the waters of the
Rhine has remained relatively stable – even though agricultural
production in the Rhine countries has been in decline[25] – exemplifyng
the uncooperative behaviour of the agricultural sector.

  In conclusion, the various public and private organizations involved in the protection of the Rhine have made widely divergent
contributions to this transnational issue. The large firms in the basin
have probably made the largest contributions, followed by domestic
policy-makers (including city officials). International policy-makers
come in at third place as only after 1987 have their efforts significantly contributed to the restoration of the Rhine. The agricultural
sector has been the least helpful. How to explain the differences
between these different cooperative stances constitutes the empirical
puzzle of this chapter.

Possible explanations

This section will scan various analyses and frameworks for possible clues to the puzzle sketched above, starting with a discussion of the usefulness of neoliberalism and neorealism. Then the extent to which the 'common pool resources' framework developed by Elinor Ostrom is applicable in the Rhine case will be discussed, following which several ideas that have been brought up in the emerging literature on the role of firms in global environmental politics will be considered. Finally a recent explanation of why the industries along the Rhine have made a significant contribution to the restoration of the river basin will be presented and criticized. The conclusion to be drawn is that these approaches and analyses do not add up to a comprehensive explanation of the puzzle of this chapter.

Neorealism and neoliberalism

Two influential, general theories of international cooperation are still neorealism and neoliberalism.[26] Neorealism posits the improbability of extensive and enduring international cooperation.[27] It argues that the anarchical aspect of the international system (i.e. the absence of an overarching authority which has monopolized the legitimate use of weapons) has forced states into a self-help system in which states have to be concerned with their own security and survival. As neorealism often assumes that it is difficult to distinguish friendly states from foes, or that it is impossible to foresee whether present friends will not turn into enemies in the future, states will be concerned with gaining a preponderance of military power over other states. Two consequences are said to follow from this. First, as all states are trying to gain a preponderance of power, a balance of power between states will often emerge. Second, international cooperation is predicted to be limited. The concern of states with their own survival will lead them to focus on the relative gains from international cooperation. States will be wary of engaging in international cooperation when this will increase the military capabilities of other states more than its own military power. As many forms of possible international cooperation would distribute absolute gains differently, international cooperation is often expected to be thwarted. This holds not only for strictly military issues (such as the formation of alliances), but also for many forms of economic coordination, as the military capabilities of a state are often thought to be dependent on the economic resources of the state.

Neoliberalism (or neoliberal institutionalism) holds that more extensive and lasting international cooperation is possible.[28] It derives many arguments for this from the assumption that information is costly for states. In the modern, complex world, so the reasoning goes, governments find it difficult and costly to assess which precise interests they have and what the intentions of their foreign counterparts are. International institutions can provide this information, thereby making international cooperation possible. For instance, international institutions can monitor the implementation of treaties by all parties, making shirking more difficult. International institutions also lengthen the time horizons of international actors. Thus it becomes possible for states to build up trust between each other. As a consequence, it becomes less important for states to be concerned with relative gains, and allows them to focus on realizing the absolute gains of international cooperation. International institutions can also link various different international issue-areas, thereby facilitating trade-offs in international negotiations. Furthermore, international institutions may spread the (scientific and bureaucratic) knowledge that is necessary for the implementation of international agreements, or the needs for knowledge among government officials may empower a transnationally organized community of scientists who not only have expert knowledge but are also less concerned with the relative gains from cooperation and care more about the absolute gains. Last, neoliberalism argues that once international institutions have been established (for instance by a former hegemon) states will find it very costly to abandon them.

Neorealism and neoliberalism have often been presented as opposites. However, it has also been noted that they share many assumptions,[29] and these commonalities make it possible to criticize both approaches simultaneously. Many critiques of both approaches have in fact been offered, but here two shortcomings of neorealism and neoliberalism are highlighted which make it impossible to understand the environmental protection of the Rhine on the basis of either approach.

First, both traditions are state centric. They present states as the main actors in the world system, sometimes even as the only significant actors. Very little attention is paid to the contributions that non-state actors may make to international cooperation. Neoliberalism sometimes considers the influence that citizens' groups and scientific experts can have on the goals and deeds of states, but almost no attention is given to how the actions of non-governmental organizations (such as firms, unions, churches or environmental groups) can *directly* contribute to the solution of a transnational issue. In being state-centric, both approaches make two implicit assumptions: (a) government regulation is needed for the solution of international issues; (b) government regulation is sufficient for dealing with international problems. As Oran Young has argued, neither assumption necessarily holds for each and every transnational problem, especially not for international environmental issues.[30] It may happen that non-governmental actors will feel responsible for the existence of crossboundary problems, and will take action to solve these problems. Moreover, governments may not have the means to induce or force domestic actors to contribute to an international agreement.[31] The two assumptions are clearly untenable in the case of the ecological revival of the Rhine. The chemical industry has taken voluntary measures to protect the environment of the river while the agricultural sector has largely ignored domestic and international legislation aimed at limiting the discharge of nitrates into waterways.

Second, the neorealist–neoliberal debate distinguishes between only two possible attitudes to international interaction: a cooperative stance, and an uncooperative one. It is expected that at a given point in time all international actors share the same degree of (non-)cooperativeness. In other words, the contenders in this debate assume that all international actors are either equally cooperative or equally uncooperative. Perhaps this may be a more reasonable position to take when the assumption is made that only states matter in international relations. But as soon as the vital importance of non-state actors is acknowledged, this dichotomy becomes too limited. Various organizations may, at the same time, contribute to the solution of a transnational problem in different degrees. With regard to the international issue of the environmental protection of the Rhine, the chemical industry and domestic policy-makers have been cooperative, the international policy-makers were rather uncooperative until 1987, while the agricultural sector has been extremely uncooperative.

In sum, neither neorealism nor neoliberalism seem useful tools with which to understand the environmental protection of the Rhine.

Common pool resources

A more promising approach with which to understand the behaviour of the Rhine actors seems the 'common pool resources' model developed by Elinor Ostrom and colleagues.[32] This model looks more promising as it focuses on self-regulation by private actors – a conspicuous feature of the Rhine regime. Ostrom and her collaborators have set up a database of common pool resources (CPRs), which is any resource with both a low jointness and a low exclusiveness of consumption.[33] Examples are fisheries, meadows and irrigation waters. The waters of the Rhine can also be viewed as a common pool resource. Ostrom's interests concern the conditions under which common pool resources can be regulated without governmental intervention. From their database Ostrom et al. have induced six characteristics of common pool resources that make self-regulation by private actors more likely.[34] Unfortunately, not a single one of these features has prevailed in the Rhine regime.

The first condition highlighted by Ostrom and colleagues concerns the presence of boundary rules. Self-organized CPRs are characterized by effective rules specifying who can and who cannot appropriate from a CPR. In the Rhine regime such rules have of course been absent. One could argue that the requirement to get a pollution permit has functioned as a boundary rule: only those firms and cities have been allowed to discharge their wastewater into the Rhine that have acquired a permit to do so. But this is not a boundary rule that has been agreed upon by the private actors themselves (which is the focus of Ostrom's work).

The second feature of successful CPRs distinguished by Ostrom is the presence of authority rules. These are rules that determine the flow of resource units through the allocation of space, time and technological capabilities. Such rules have also not prevailed in the Rhine regime. (One should remember that Ostrom is solely concerned with rules decided upon by private actors, not by government organizations.) The voluntary actions undertaken by firms and municipalities have not been subject to collectively developed rules in any significant way. Of course there has been some coordination between firms and cities in the Rhine basin. For instance, in Germany the chemical firms coordinated their water protection policies through the German Association of the Chemical Industry (VCI). Furthermore, the Abwassertechnische Vereinigung (the professional association of German wastewater specialists) has defined standards that have been taken up by firms and cities,[35] while in the Netherlands, environmental managers from companies located in the Rhine catchment area have exchanged information and agreed on common policy positions in informal working groups.[36] But in the main, the decisions especially by firms to invest in water protection do not seem to have been influenced by environmental protection measures of other actors in the Rhine valley. During the interviews, business representatives reported that environmental policies within their firms had been conceived independently from the policies followed by other corporations. In other words, Rhine firms have decided to 'do the right thing' on their own, in the absence of any commonly agreed-upon authority rules.

Ostrom et al. also found that self-organized CPRs have developed active forms of monitoring and sanctioning. Again, this is not applicable to the private actors in the Rhine regime. As the firms and cities have never shown any interest in developing allocation rules in the first place, there has been no need or possibility for them to engage in the monitoring or sanctioning of other private organizations.

The same applies for the fourth condition mentioned by the CPR model of Ostrom: the absence of grim trigger strategies in case of non-compliance. Again, as the private actors in the Rhine regime have not wanted to develop any common rules, it is pointless to talk of the absence or presence of 'grim trigger strategies' in the Rhine regime.

The fifth feature brought up by Ostrom and colleagues is the size of a CPR. They argue that smaller CPRs are more likely to be self-organized than larger ones. Again, this does not shed any light on the case of the Rhine regime. Compared to the CPRs studied by Ostrom et. al., the number of organizations that have been involved in the Rhine regime has been very large.

The last characteristic of successfully self-organized CPRs concerns the physical characteristics of the resource units. Ostrom and her fellow researchers distinguish between four types of common pool resources on the basis of two physical attributes. The first of these attributes is the stationarity of the resource units (i.e. the degree to which the resource units can be transported). The second physical variable concerns storage, or the extent to which resource units can be stored. By assigning two values to each of these two variables Ostrom et al. arrive at four types of CPR. The type of CPR that is least likely to be self-organized is the situation in which storage of resource units is not available and the resource units are non-stationary. Regrettably, this is exactly the type of CPR that the Rhine represents. So the physical characteristics of the resource units of a successful CPR that are outlined in Ostrom's model are also unable to provide any insight into the voluntary actions undertaken by cities and firms in the Rhine regime.

The basic flaw in Ostrom's model is its reliance on the idea that private actors can only regulate a common pool resource through *collective* action. The Rhine regime has to a large degree been based on self-regulation that consisted of *individual* action. This form of uncoordinated self-regulation is not captured by Ostrom's approach.

Reflections on the role of firms in global environmental politics

In the last couple of years a few scattered publications have acknowledged that business corporations can play a supportive role in the design and implementation of (inter)national environmental policies.[37] Two findings have been made. First, it has been stated that multinationals in particular have increasingly sought to participate in environmental policy debates, sometimes endorsing strict regulations. Several reasons have been suggested in the literature for the increasing interest among firms in the design of environmental policies. It has been noted that firms often have different policy views than government agencies and environmental groups. Corporations usually find the traditional command-and-control policies too static and inefficient. Instead of being told not only which environmental aims to target, but also how to reach those aims, companies would prefer environmental policies that only spell out the goals to be achieved. Each affected company can then decide for itself what is the most efficient way of meeting the policy targets. Furthermore, firms seem to prefer environmental policies that make use of the market mechanism. One can think of tradable pollution permits for air and water pollution. According to the literature on the role of firms in global environmental politics, these different policy views are one impetus for corporations to get involved in environmental policy processes.[38] Business leaders feel that the views and preferences of environmental groups have thus far been privileged by governments and international organizations.[39] Another motivation that has been pointed out in the literature concerns the competitive effect of strict environmental regulations. It has been suggested that large companies sometimes favour stringent environmental legislation as it can have a disruptive effect on small companies. Supposedly, large companies find it easier to comply with strict environmental legislation than smaller firms, due to the superior administrative, financial and technological resources of the former.[40]

Much evidence for the above statements in the case of the environmental protection of the Rhine has not been found. Firms in the Rhine basin do not seem to have increased their attempts to influence government policies. Moreover, multinationals have not tried to squeeze smaller companies out of markets by insisting on stringent environmental legislation. If anything, large firms in the Rhine valley have tended to argue that governmental environmental policies have been too strict. This is quite paradoxical considering the many voluntary measures that Rhine firms have taken. An attempt to explain this paradox will be made later in the chapter. Moreover, in Germany, the giant chemical firms have actually helped smaller chemical enterprises in reaching environmental standards through a programme set up by the German Association of the Chemical Industry (VCI).[41]

The emerging literature on the role of firms in global environmental politics has on occasion conceded that firms have sometimes gone beyond compliance with governmental policies for the environment. This is the second main finding of these publications that is to be highlighted here. Again various reasons have been offered. It has been argued that firms have taken voluntary environmental measures in order to create a 'green' corporate image and thereby stimulate sales. In this view, corporations want to maximize the (present value of) their long-run profits rather than their profits in the short term. It has also been argued that multinationals routinely develop technology that is in compliance with the strictest standards of all the countries in which they produce. As a result, these multinationals are 'beyond compliance' in the remaining countries.[42] It has been said as well that corporations sometimes engage in environmental protection simply because it saves costs. Companies can sometimes reduce their pollution by making production processes more efficient.[43]

Again the present research on the restoration of the Rhine does not corroborate the above observations. In particular, the argument that companies aim to boost sales by building a 'green corporate image' does not seem to hold for Rhine companies. It has not been possible to find any sustained attempt by Rhine firms to publicize their investments in water and air protection. One would expect firms to attempt to draw attention to their protection efforts if these efforts were undertaken to stimulate sales but there seems to be no evidence of such attempts. The received impression is that the 'public at large' is not aware of the extent to which the Rhine has been cleaned up over the last 25 years. The 'sewer of Europe' image still seems to be very much associated with the Rhine. As no survey has been undertaken on the topic, this remains simply an observation, but in the course of the present research, many people have politely enquired after this topic and a common reaction has been 'Isn't that river very dirty?' If it had been the aim of Rhine firms to boost their sales by investing in water and air protection, then their efforts must rank among the bigger of marketing failures.

The assumption that firms take environmental measures to cut production costs applies better to the Rhine regime. During interviews, environmental managers of multinationals in the Rhine valley appeared very much aware of the opportunities to save costs and resources by taking environmental measures. Still, it is difficult to believe that these savings have to any significant degree compensated for the tens of billions of deutschmarks that Rhine corporations have spent in constructing and operating sewage treatment plants.

The overall problem which arises with this emerging literature on the role of firms in global environmental politics is a moral one. In these publications, the willingness of firms to invest in environmental protection cannot be accepted at face value. Some financial interest must remain hidden in the background. Firms only support environmental policies to hurt competitors, to boost sales, to cut costs or to avoid future regulation, and so on. In this view, corporate leaders do not invest in environmental protection simply because they feel that it is the right thing to do. Perhaps some will find all of this simply common sense: firms are only interested in profits and shareholder value. But it is important to realize the full import of this assertion: that business executives are willing to poison people (including themselves, their families, friends and employees) as well as animals and plants just to increase sales and profits, and only when corporate leaders see a way of increasing output and shareholder value by not poisoning human and animal life will they voluntarily stop such practices. Such a simplistic view is hard to accept. Corporate leaders are indeed willing to devote resources to environmental protection, provided that they are convinced of the harm that production processes lead to. The problem is that business leaders are often less readily swayed by evidence pointing to the harmful environmental effects of production processes than, for instance, are environmental groups. An advantage of grid-group theory is that it restores morality to actors. Each actor is seen as inherently moral, but the point is that actors follow different moralities. The theory states that actors who adhere to one morality will try to discredit the moralities followed by others in order to boost the acceptance of their own morality.

An analysis of the Rhine regime

Thomas Bernauer and Peter Moser have also compared the governmental and private efforts to clean up the Rhine,[44] and their findings support the puzzle set out above. The authors agree that the intergovernmental cooperation on this topic has by and large been ineffective, certainly before the adoption of the Rhine Action Programme in 1987. They also point out that national water laws and policies went beyond international agreements, and that the anti-pollution measures taken by industries in the river basin went even further than domestic laws. However, their research omits pollution of the Rhine by agricultural enterprises.

In their analysis, Bernauer and Moser attempt to explain the voluntary investments in environmental protection made by private firms in the Rhine watershed. They advance three reasons, although it is contended that these three arguments do not add up to a full explanation. First, the authors point to the influence of 'the shadow of future legislation'. The argument is that although no strict domestic, international or European water policies were in place when the chemical firms decided to invest in water protection, they acted on the belief that such legislation was forthcoming in the future. In other words, the Rhine companies merely anticipated future legislation. In the present research there was some evidence supporting this position. At the beginning of the 1970s, German political parties and governmental officials were discussing the pros and cons of high levies for polluters. In the end these proposals were not adopted, but the discussion partly induced at least one major chemical company in the Rhine basin to start investing in water protection. Yet, the strength of the argument should not be overestimated, as it poses several problems. The argument overlooks the fact that the development of domestic, international as well as European water protection policies has been shock-driven. International as well as European negotiations on water protection

measures remained deadlocked until the shock of the Sandoz incident in 1986. German water protection laws were much strengthened
only after the widely publicized seal deaths in the Waddensea in
1985. Therefore, to argue that the investment decisions taken by
Rhine companies anticipated future legislation is also to say that
the leaders of these enterprises were able to foresee how stalled
political processes would become fluid again as a result of acci-
dents. However competent these business leaders may have been,
such powers of prescience must have been beyond their, or anyone
else's, capabilities. In addition, the 'future legislation' argument leaves
untouched the question of why the chemical firms developed revolutionary techniques for wastewater treatment that made reductions
in discharges of pollutants feasible that had previously been thought
to be impossible. The worst the firms could have expected from
future legislation was the obligation to use 'best available technical
means'. So, if they had been reacting to future legislation, they
could have limited themselves to using existing technology for
wastewater treatment instead of spending a lot on research and
development in a quest to develop more much effective technology. It could even be argued that in this way the chemical industries
led and directed future legislation instead of the other way around.
Furthermore, the 'future legislation' thesis leaves unexplained why
the agricultural sector has not anticipated future legislation. In fact,
the agricultural firms have ignored not only future legislation but
also laws adopted in the past. Last, the voluntary measures taken
by companies in the Rhine watershed seem simply too extensive
for the 'future legislation' argument to hold. As shown above, even
by the mid-1980s the Rhine enterprises had reduced their pollution
by more than was required by the quite strict standards that were
formulated in the 1990s. No imaginable legislation would have demanded such massive investments by the Rhine companies.

The second argument offered by Bernauer and Moser concerns
the 'environmental consciousness' of the citizens within the Rhine
countries. This argument has two versions. The first is that, given
the growing environmental concern among the citizens of the riparian countries since the 1960s, it would have been bad publicity
for companies not to engage in environmental protection. Instead,
by investing heavily in water and air protection, firms could build
up an environmentally friendly reputation, thus safeguarding their
sales. As argued above, the present research does not yield much
evidence to support this assumption. The second version of the

'environmental consciousness' argument states that the growing environmental awareness not only affected the consumers of the products sold by chemical firms, but also influenced the people working in these enterprises. In particular, Bernauer and Moser point out that the large chemical concerns along the Rhine have established extensive environmental departments that have hired leading experts in the field of wastewater treatment. The authors argue that these environmental departments have probably been able to persuade top management to invest in protection measures. It cannot be denied that these departments have tried to do so. However, their influence on the highest management level should not be exaggerated. A decision to build sewage treatment plants is not a creeping policy change that goes undetected by corporate leaders. Such a decision commits billions of deutschmarks that could also be invested otherwise. As a consequence, these decisions are only taken after careful deliberation. Moreover, this argument does not address why business leaders decided to set up large environmental departments in the first place.

The last proposition offered by Bernauer and Moser posits that the extensive financial resources of the chemical companies in the Rhine basin have enabled these firms to make large investments in environmental protection. They point to the large profits that the chemical industries in particular enjoyed throughout the 1970s and 1980s. This is a convincing argument. The large chemical concerns have had the financial means to set up large environment divisions, develop new wastewater technology, as well as implement and monitor expensive environmental policies. The agricultural sector has consisted of much smaller firms, with fewer financial, technological and administrative resources, and even government agencies have been short of funds. Command-and-control policies are expensive to implement fully, and the greater resources of (especially the large chemical) firms in the Rhine valley make up one factor that helps explain why the industrial sector has made more extensive contributions to the clean-up of the river than other actors. Still it cannot have been the only factor at work. Besides having the resources for environmental protection, actors must also want to spend these resources on pollution prevention. Moreover, the financial argument cannot explain the following paradox. Several agricultural experts have argued that, as a rule, farmers are using too much fertilizer on their lands, and have cited scientific reports that show that farmers would obtain a similar harvest by using half the amount of fertilizer that they are currently putting on the land. If farmers were to do this, they would receive a similar income but with lower costs and less pollution. However, until now, very few farmers have been willing to try out this strategy.[45] The financial argument fails to address this paradox. What is needed to explain it is some understanding of how harsh economic conditions affect the mindsets of actors.

Another economic argument can be added to the one offered by Bernauer and Moser. According to many of the government officials interviewed, the pollution of the Rhine by industrial sources has also declined because the economic crises of the 1970s and 1980s forced many companies to reduce their output or even to stop production altogether. This argument does not apply so much to the chemical sector which on the whole continued to be profitable throughout this period, but it certainly is true for the steel companies as well as the coal mines that were situated in the Rhine basin. The steel firms in the Ruhr area were especially hard hit by the economic crises, and this has no doubt also contributed to the reduction of pollution in the Rhine. But again this explanation needs to be put in perspective. The chemical industry in the Rhine basin has not declined in size and profitability, and generally speaking it is this sector which is responsible for most of the pollution in an industrialized water basin. In addition, while the steel industry in the German Ruhr area declined by some 30 per cent between the mid-1970s and mid-1980s[46] – in economic terms a huge decrease – it is still too small a decrease to fully account for the pollution reduction in the Rhine basin of some 80 to 90 per cent. Last, the argument cuts both ways. On the one hand, a decline in economic activity and profitability leads to a decrease in the production of the amount of pollutants. On the other hand, a decline in profitability also leaves firms with less money to spend on keeping pollutants out of their wastewater. In the end, the net effect of economic decline on the discharge of pollutants remains rather unclear.

In conclusion, the host of theoretical frameworks and specific analyses throws up just one argument that helps explain the puzzle of this chapter: a straightforward financial one. This single argument certainly does not provide a full account, not even when combined with the 'economic decline' thesis. In the next section, it will be argued that the cultural regime analysis developed in Chapter 3 provides a better explanation.

A cultural explanation

Besides financial constraints, the ways of life adhered to within firms, farms, ministries and environmental groups have also influenced the contributions that these actors have made to the clean-up of the Rhine. Establishing which ways of life have prevailed within these organizations is difficult, in particular because a great many governmental, international and private organizations have been involved in the Rhine regime since at least the mid-1960s. In 1996 and 1997, interviews were conducted with 58 officials from 46 different organizations, including firms both national and local, European policy-makers, agricultural associations, environmental groups and international commissions (see Appendix A). In addition, a review was undertaken of the academic literature on the environmental protection of the Rhine and of publications by stakeholders. The ways of life of the various groups of private and public actors were established as follows: only when a large majority of the persons working for one type of organization expressed a certain norm or belief was it accepted that this norm or belief had prevailed within this type of organization. The limits of this empirical research need to be emphasized from the outset. Often it was only possible to speak with a single representative from an organization (such as a particular firm) at a single point in time, and this may have limited the insight from the research in two ways. First, the values and views of the persons spoken to may not have overlapped fully with the views held by other persons working for the same organization. Second, at other points in time, interviewees may have had different ideas. The first problem was eased somewhat by several factors. Sometimes, it was possible to speak with various officials from the same organization. Also, by grouping organizations together (such as industrial enterprises, domestic and international authorities, and the agricultural sector) the problem of having spoken with only one person from a particular organization was somewhat ameliorated. For every such *type* of organization it was posssible to interview a number of officials. In addition, interviews were often held with the official representatives of these groups of organizations, for instance representatives from the CEFIC and VCI which are, respectively, the associations representing European and German chemical firms. By definition these representatives are supposed to speak for all member organizations. The second problem was eased by the fact that the majority of officials interviewed had been involved in the clean-up of the Rhine for a long period of time.

Below are described the ways of life to which actors of the Rhine regime adhere. Also shown is how these particular ways of life have affected the contributions that actors have made to the restoration of the Rhine. In this section, the way of life espoused by (most) environmental groups has also been included, although such groups have not contributed directly to the clean-up of the Rhine through investment decisions or legal measures. Their influence has been more indirect, by informing the press, by demonstrating near the sites of firms and governmental meetings and by discussing issues with other organizations involved in the regime. Thus it appears that the actions of environmental groups have not been very influential in the clean-up of the Rhine, and while they have certainly raised the awareness of everyone involved, they do not seem to have influenced strongly either investment decisions by firms or government policies. Nevertheless, environmental groups have been included below, as the rationale that has prevailed among environmental organizations forms a revealing contrast with that of other actors.

Large industrial firms[47]

The principles and norms, as well as rules and procedures, that have been followed by large enterprises in the Rhine basin have tended to be individualistic.

Principles and norms

The large companies in the Rhine catchment area[48] have shown a preference for a minimum of intergovernmental governance of the Rhine. For instance, during recent years they have seen no further need to continue the existence of the ICPR. They have acknowledged that the ICPR may still be useful in restoring habitats in the river basin, but have also felt that continued involvement of the ICPR with the issue of water pollution would be a waste of resources. Consequently, abolition of the ICPR would not be felt as a great loss by the large firms.

Cynics could easily explain this stance. The Rhine corporations would be happy to see the ICPR go, as this would reduce pressure on firms to make costly investments in water protection. But this argument does not make much sense in the light of the many voluntary contributions that the firms in the Rhine basin have made to the restoration of the river. Firms have actually appreciated the activities of the ICPR, but they see no need to continue with an international commission whose job they feel has been done. This reflects an individualistic preference for organizing international governance. Instead of ICPR agreements, large Rhine firms have preferred EU regulation of water protection. This would create a more level playing field for companies. At the same time, however, corporations have been wary of 'European bureaucracy'.[49] They have favoured a system of simple and clear quality aims for surface waters in the whole of the European Union. This would keep international governance to a minimum, while not distorting competition between European companies.

Trust between the large corporations in the Rhine basin has abounded. Firms have trusted each other to take similarly expensive environmental measures. In Germany, chemical companies have coordinated their protection efforts in the German Association of the Chemical Industry (VCI).[50] In the Netherlands, chemical and petrochemical concerns have reached informal understandings in working groups and professional associations. The large firms in the Rhine valley do not appear to have been overly worried that other firms would try to shirk their responsibilities.

A last principle/norm concerns the myth of nature that has tended to prevail within Rhine companies. Myths of nature have not been included among the principles of the four international ways of life set out in Chapter 3, which have only incorporated principles and norms concerning transnational issues. By contrast, a specific myth of nature applies to both domestic and transboundary ecosystems. However, in Chapter 2 it was shown that the followers of the four alternative ways of life also believe in different myths of nature. Individualists believe nature to be resilient. To them, ecosystems are too robust to be much affected by human activities, and it is this myth of nature that comes close to the ideas that have tended to be held within large companies in the Rhine basin. Rüdig and Kraemer have found that before the 1970s:

> Public waterways were largely seen as a 'free good' by [German] industry, and effective regulation was successfully resisted. Industrial pollution was also seen as essentially 'harmless', and the great rivers like the Rhine were ascribed great natural cleaning powers.[51]

A similarly optimistic view of nature has also reigned within Rhine enterprises in later years. Firms have tended to consider a smaller number of chemicals as harmful than have ministries and environmental groups. Likewise, companies have not been overly worried about the presence of very small amounts of toxins in the waters of the Rhine. In their view, such small amounts can do no harm. Environmental groups, however, have tended towards the opposite opinion. At present, Rhine firms tend to believe that the river system is fully cleaned up. By contrast, although government agencies and environmental organizations are willing to acknowledge that many improvements have been realized, as a rule they do not consider the ecosystems of the Rhine to be fully revived.

Given their view of nature as essentially resilient, how is it that the Rhine firms have invested massively in water protection? In grid-group theory, reality is malleable, or 'multi-interpretable' (to use a cumbersome term), but not to any degree. At times, reality vetoes certain interpretations of the world. In the late 1960s and early 1970s, it was evident 'for all to see' (and smell) that the Rhine was highly polluted. To rebuff the most ardent supporters of the thesis that the river's natural cleaning powers were endless, there was the incident in June 1971 when a large section of the Rhine no longer contained any oxygen. So, from the end of the 1960s onwards it had become undeniable even by Rhine companies that the river was highly polluted. The view of nature that has by and large been held within Rhine companies was only more robust compared to the views of nature adhered to within other organizations.

The individualistic principles and norms that have been favoured within large companies in the Rhine basin offer a basis for explaining a paradox. At times, companies in the Rhine watershed have protested against the adoption of effluent limits and water quality standards by government agencies. This is quite baffling, given that at the same time, the same firms were making extensive voluntary investments in sewage treatment. It is even possible that these contradictory developments have at times wrong-footed ministries and environmental groups. Indeed, the latter appear to have focused at times more on the rhetoric of Rhine companies, which has tended to be much less 'environmentally friendly' than their actual practices. Grid-group theory helps to clarify this paradox. Firms have protested against governmental prescriptions, as extensive governmental regulation does not square with their preference for limited governance. Quite apart from their views on the pollution of the

Rhine, they have reacted against the central regulation of production activities that comes as part of governmental efforts to clean up the environment.

Rules and procedures

The rules and procedures that have been followed by large enterprises in the Rhine valley have also tended to be individualistic. Firms have been opposed to the command-and-control strategies followed by governments, viewing the simultaneous prescription of goals and means by governments as inefficient and too static. In their view, such policies do not stimulate technological innovation. They have also argued that governments do not have the detailed knowledge that is required for imposing technical means on industries. Governments should decide on environmental goals, and firms should be free to decide on how to reach those standards. In addition, according to Rhine corporations, chemical substances should only be regulated on the basis of scientific certainty. They have tended to reject the precautionary principle. Large enterprises have also stressed the need for taking efficient water protection measures. They have argued that increasingly strict effluent limits lead to ever smaller environmental benefits at ever greater cost. Water protection measures should be taken where the greatest environmental benefits can be had for the least amount of costs. Rhine firms have pointed out that such measures could especially be taken in Central and Eastern Europe. In their management approach to water protection, the large firms in the Rhine basin have been risk-takers. They first set themselves environmental goals and then proceeded to develop revolutionary technology to reach those aims. This practice is quite different from the 'best available means' approach that has sometimes been favoured by governments. In the latter approach, environmental goals are made dependent on existing means, instead of the other way around.

It can be concluded that the principles and norms, as well as the rules and procedures, followed by the large industrial concerns in the Rhine catchment area have tended to be individualistic. This conclusion is somewhat surprising. It might be expected that the environmental discourse of these enterprises would be more infused with hierarchical notions because of the internal organization of the firms taking part in the research. All of these firms (BASF, Bayer,

Hoechst, Shell, Ciba-Geigy, Sandoz) are vast and highly complex organizations, and have hired leading experts in water protection for their environmental departments. These hierarchical elements in the internal organization of the companies would lead one to expect similarly hierarchical elements in their environmental discourse, but this does not appear to have been the case. Two explanations for this may be put forward. First, the *external* environment of these firms resembles much more an individualistic social structure. After all, these firms are operating in competitive world markets. Second, the *internal* organization of these firms also incorporates individualistic traits – besides the hierarchical elements mentioned above. For example, Hoechst AG in Frankfurt has been subdivided into 140 more or less independent profit centres. This has been done in an effort to keep production processes as efficient as possible, as well as to curb bureaucracy within the company. The other enterprises have followed similar strategies. In all, it can still therefore be argued that the social setting in which large Rhine companies have operated is more individualistic than the social environments of other stakeholders involved, such as government agencies.

But how can the individualistic way of life espoused by the large companies in the Rhine basin explain their extensive contributions to the ecological revival of the river? The trust that the companies have had in each other has allowed them to move speedily ahead with their own investments in water protection. Unlike the national delegations to the ICPR, these companies have not wasted much time in disagreeing with each other. Furthermore, their dynamic and risky approach to solving pollution problems has also been important. The firms adopted ambitious environmental standards and then designed new sewage treatment technology to reach these standards. At the outset of these programmes, it was not at all clear that such technology could be invented, but the large enterprises were willing to take the risks. This risk-taking approach contrasted with the more careful policies of government organizations, that relied at best on the notion of 'best *available* technology'. Above all, seeing large enterprises as individualistic actors restores them as morally responsible actors. As individualists, they have tended to view the Rhine as a robust ecosystem. As such, they have disagreed with other stakeholders over the need to take environmental measures. But when it had become plain in the late 1960s that the limits of the Rhine's resilience had been reached, the large firms in the river basin were not at all opposed to investing in environmental protection. For them, it was simply another job to be done.

Environmental organizations[52]

The principles and norms as well as rules and procedures enacted by most, but not all, environmental organizations involved in the Rhine regime have tended to be egalitarian. The main exception to the egalitarian discourse of the environmental groups is formed by the actions and viewpoints of the more 'moderate' World Wide Fund for Nature – Germany.

Principles and norms

Environmental groups concerned with the protection of the Rhine have tended to view nature as fragile. In grid-group theory, this is the egalitarian myth of nature. Environmental groups have tended to believe that each chemical substance entering the Rhine watershed can have harmful effects on the flora and fauna in the river. In other words, environmental organizations have regarded many more chemical substances as toxic than firms have. In addition, environmental groups have argued that even very small amounts of pollutants in the Rhine may have detrimental health effects. Again, firms have disagreed. According to companies, very low concentrations of pollutants in the Rhine cannot be harmful. As a result of their particular view of nature, environmental groups in the Rhine valley have offered a different assessment of the clean-up of the river basin. They have acknowledged that significant progress has been made, especially in the reduction of pollution by traditional pollutants, but they have also pointed out that the chemical industry has emitted thousands of different chemical substances into the Rhine. Only a fraction of these substances have been subjected to toxicological research and governmental regulation. Environmental organizations have suspected that many of the remaining chemical substances are also hazardous. Therefore, their evaluation of the restoration of the Rhine has been less enthusiastic than that of firms and governmental organizations.

Even in the Rhine watershed, environmental organizations have been distrustful with regard to industrial firms and authorities. They have abided by the idea that corporations are only interested in profits. In their view, only strict governmental regulation and public pressure can force firms into respecting the environment.

Environmental organizations have also tended to believe that a full restoration of the Rhine basin can only come about through a fundamental transformation of current political and social conditions. They have clung to the view that capitalism and the profit motive are ultimately incompatible with a harmonious relationship with nature, and have argued that existing social inequalities have perpetuated an ecological crisis. In the view of environmental organizations, the currently rampant 'ego-centrism' should be replaced with 'eco-centrism'. For the Rhine basin, environmental groups have promoted more open decision-making structures. For instance, they have argued that the ICPR would benefit from much more public participation. These preferences for specific forms of governance are essentially egalitarian.

How have the opinions of WWF – Germany differed from the above standpoints? WWF has self-consciously abstained from a radical critique of capitalism and prevailing power divisions. Also, the organization has had more faith in the good intentions of firms and the capabilities of governments. The rules and procedures that have been advocated by WWF overlap more with the rules and procedures that have been preferred by other environmental organizations in the Rhine catchment area.

Rules and procedures

Rules and procedures favoured by environmental groups have also tended to be egalitarian and a very strict application of the precautionary principle has been urged by these groups. As the environmental organizations in the Rhine basin have suspected many chemicals to be highly toxic, they have preferred a system in which chemical substances first undergo toxicological tests before being allowed into the water. In their view, only if scientific proof exists that a chemical substance does not harm the environment can permission be given for its discharge into the water. Such a strict application of the precautionary principle would cripple the present-day chemical industry in the Rhine watershed. Each year, the chemical firms in the Rhine area invent many new chemical substances. If each of these had to be tested in toxicological research, the ongoing production processes of the chemical firms would be severely hampered.

Nature redevelopment in the Rhine watershed has also been a main concern of environmental groups. Since the early 1980s environmental organizations have been highlighting the harmful effects of both the canalization of the Rhine and the expansion of human settlements along the river. In their view, instead of controlling the river through engineering works (dams, weirs, canals, dikes), people should learn how to adapt to the needs of the flora and fauna in the Rhine watershed.

The environmental groups in the Rhine basin have not greatly influenced decisions taken by firms or government agencies. Examples of attempts to influence decision-making concerning the Rhine are the court cases against the Mines de Potasse d'Alsace that were co-organized by the Dutch Stichting Reinwater, and the water tribunal organized by a number of environmental groups in 1983. Environmental groups also staged demonstrations in the days following the Sandoz accident. Despite these activities environmental organizations in the Rhine basin have not exerted much *direct* influence on the protection of the Rhine. First, environmental organizations in the Rhine countries have not concentrated on water protection. Instead, they have focused more on the risks of nuclear energy and air pollution. To the extent that they have been concerned with water pollution, they have tended to zoom in on groundwater pollution by the agricultural sector.[53] Greenpeace lost interest in the protection of the Rhine when it discovered that the watershed was actually quite clean. Moreover, the other stakeholders in the Rhine area have remained rather closed to environmental groups. For instance, it was not until 1996 that the ICPR began to consult NGOs. National authorities have met much more often with environmental organizations, but have never let themselves be dictated to by these groups. Firms appear to have had few contacts with NGOs. Environmental movements have also been unable to exert an *indirect* influence on especially the investment decisions made by firms. The waxing and waning of the environmental movements within the Rhine countries has been unrelated to the timing of investments in water protection in the Rhine watershed. Support for environmental groups and environmental protection in the Rhine countries took off in the mid-1970s, but the large companies in the basin had already decided in the mid to late 1960s to take environmental measures, thus preceding the rise of the environmental movement, while during interviews, business spokesmen declared that the water protection decisions of their firms had not been influenced by the actions of the environmental groups. In conclusion, it does not seem that the mostly egalitarian environmental groups in the Rhine basin have been able to directly or indirectly affect the environmental protection of the river to any significant degree.

Governmental agencies

The public authorities that have been concerned with the protection of the Rhine encompass many layers of government: municipal (responsible for the construction of sewage treatment plants in cities), regional (the Swiss cantons, the French *agences de l'eau*, the German *Länder*), national (the Ministries of Environment, Transport, Foreign Affairs, Economic Affairs), international (the International Commission for the Protection of the Rhine, the Central Commission for the Navigation of the Rhine) and European (the European Union).[54] Most governmental agencies have not only been responsible for domestic water protection policies, but have also been active in the international negotiations concerning the environment of the Rhine that have been held under the auspices of the ICPR. The Deutsche Kommission zur Reinhaltung des Rheins (in which the federal ministries and the *Länder* try to coordinate their policies on a voluntary basis) has represented Germany at international conferences. The position of the French state is determined at the central level, but the local *agence de l'eau Rhin–Meuse* is represented in the French delegation to international meetings. In Switzerland, the Bundesministerium coordinates its position with the cantons before the international negotiations begin.

Chapter 4 described the principles, norms, rules and procedures to which the government agencies involved in the protection of the Rhine have subscribed – both at the domestic and at the international level. These predispositions have tended to be hierarchical, at least when compared to the principles, norms, rules and procedures that have been followed by corporations and environmental organizations. The principles and norms of the government agencies concerned have been characterized by low levels of trust and competency struggles (especially at the international level), as well as an insistence on the primacy of governmental rule. Their rules and procedures have emphasized command-and-control approaches to water protection and prescription of technical means. Here, two further points need to be made.

First, it appears that at the end of the 1980s the government agencies involved in domestic water protection started to diverge more and more from traditional command-and-control approaches, and began to experiment with other forms of regulation. These newer forms of governance have been based on two principles: (a) more reliance on self-regulation by firms (for instance codified in covenants between companies and ministries); and (b) more consultation with environmental movements. In other words, government agencies responsible for water protection have become less heavy-handed and have sought to establish a better dialogue with both firms and environmental groups. These processes took off in all Rhine countries at around the same time,[55] and represent a partial breach with the more purely hierarchical methods that agencies favoured before the end of the 1980s.

Second, although it can generally be said that government organizations have adhered to a hierarchical way of life (especially when compared to other stakeholders in the basin), some of them have been more hierarchical than others. The Ministries of Foreign Affairs have been stalwarts of hierarchy. The discourse of other ministries has been more coloured by the ways of life followed by their constituencies. For instance, Ministries of Economic Affairs have tended to lean more towards the individualist opinions espoused by firms. During inter-ministerial consultations over the protection of the Rhine, these ministries have tried to get the precautionary principle watered down, by defending scientific certainty as the only sound basis for regulation. These ministries have also emphasized the need for a cost–benefit analysis of each proposed environmental measure, positions which resemble individualist policy views held by the companies in the Rhine basin. The same applies to the Ministries of Agriculture in the Rhine countries. Their discourse has tended more towards the fatalistic views that have been expressed by farmers' associations (see below). In sum, in general the governmental organizations involved in the restoration of the Rhine valley have tended to adhere to hierarchy, but several ministries have also taken over various elements from the ways of life followed by the private organizations within their jurisdiction.

How have these hierarchical practices and attitudes affected the contributions that government agencies have made to the environmental protection of the Rhine? As set out variously in this book, the hierarchical way of life seems to 'work better' domestically than across state boundaries. Chapter 4 argued that several elements of the hierarchical way of life followed by governmental delegations to the ICPR blocked any progress on international measures to protect the Rhine. First, there were ongoing competency struggles, as well as distrust, between the delegations. Second, there was the insistence on following all the traditional rules attached to the making of intergovernmental treaties, in which process a lot of valuable time was lost. Last, the government delegations to the ICPR tried to construct an overarching command-and-control system of water protection that sat uneasily with the command-and-control approaches that had already been developed within the countries. For these reasons, the governmental efforts to protect the Rhine through international agreements were by and large fruitless before the Sandoz incident in late 1986.

At the domestic level, the hierarchically inclined government agencies have been much more constructive. At the domestic level, hierarchical organizations feel more at home. This may primarily be because somewhat clearer lines of authority exist in the domestic realm than internationally. Among other things, this diminishes the space for competency struggles. Concretely, at the domestic level government organizations have induced both firms and cities to build sewage treatment plants. Of these two types of actors, municipalities seem to have been more under the control of the authorities than corporations. City officials stated that their water protection policies quite closely followed instructions issued by higher level government agencies.[56] Firms appear to have made their investment decisions more independently. For example, corporations in the Rhine area have independently decided to invest in the technological advancements that have made the extensive restoration of the river basin possible. Therefore, the credit for cleaning up the discharges of municipalities (which include the effluents from smaller companies) should go especially to domestic authorities.

To sum up, the government agencies concerned with the environmental protection of the Rhine have tended to espouse hierarchical principles, rules, norms and procedures in both domestic and international policy realms. As a consequence, their efforts in the domestic realm have had a more significant impact on the clean-up of the Rhine than their international attempts.

Agriculture

The agricultural sector has made the smallest contribution to the clean-up of the Rhine. One factor that helps to explain this development has already been pointed out: since the 1980s the incomes of farmers in the European Union have continuously decreased, while the profits of especially the chemical firms in the Rhine basin have steadily risen.[57] As a result, industrial firms have had more money to spend on water protection. Still, the full import of this financial argument can only be appreciated when it is seen in the light of the much larger social and political changes that have taken place in the agricultural sector in Europe since the late 1980s. Broadly speaking, since that time the agricultural sector in the Rhine countries has moved from a highly protected, hierarchical social system to a more fatalistic social system, a change reflected in the opinions and views of agriculturists in the Rhine basin. This move to fatalism, both in social relations and cultural bias, helps in understanding why farmers have not been more cooperative towards the issue of the environmental protection of the Rhine.

Before the 1990s, agricultural production in the European Union was highly protected and regulated through the Common Agricultural Policy (CAP). The incomes of farmers were sustained by export subsidies, high tariffs and sale guarantees. As a consequence, the prices for agricultural products in the EU were kept higher than world market prices. Farmers could make more money by simply increasing their production. This highly regulated system of production was supported and implemented by an 'iron triangle' consisting of the Agricultural Directorate (DG VI) of the EU, the national Ministries of Agriculture within the member states and agricultural associations. These organizations worked closely together and excluded non-agricultural organizations from the decision-making processes.[58] At the end of the 1980s it had become clear that this quintessentially hierarchical system of organizing agricultural production had to change and was challenged under the rules of the General Agreement on Tariffs and Trade (GATT). It had led to increasing environmental pollution by diffuse sources as a result of the expansion of production and the intensification of the use of farmland, and it had brought about budgetary problems as well as sky-high stockpiles of agricultural products. It had not even stopped a decline in the incomes of farmers.[59] In 1992 the Agricultural Council of the EU was ready to commit itself to a series of changes that have come to be known as the MacSharry reforms. These measures have brought the agricultural sector in the European Union closer to a fatalistic social system. The MacSharry reforms of the CAP intended to replace price controls with direct forms of income support for farmers. As such, it sought to bring EU prices for agricultural products back in line with world prices. Also, it aimed at a reduction of agricultural production within the EU. Farmers were to be paid for keeping their production and livestock below a certain maximum. These reforms were accompanied by a set of new environmental measures, the most important of which was the Agri-Environmental Regulation.[60] Under this regulation, farmers receive financial aid if they reduce their use of fertilizers and pesticides, decrease the proportion of sheep and cattle on their forage areas, as well as set aside farmland for at least 20 years (among other things). All these policy innovations seem to fit well into the hierarchical, centrally regulated production system that existed before. How then can it be said that the MacSharry reforms brought about more fatalist social relations?

In several ways. First, the MacSharry reforms have increased the rules and regulations to which farmers are subject. Today, farmers have to report many more aspects of their activities to the authorities: total production, number of cattle and sheep per forage, amounts of fertilizers and pesticides used on their lands, etc. This can clearly be seen as an increase of the grid-dimension of the social context of farmers. Second, the reforms have diminished solidarity with, and between, farmers. The MacSharry measures have substituted price controls with direct forms of income support. At first, this did not greatly affect the incomes of farmers,[61] but over time the situation will be different. Declining world market prices for agricultural products, coupled with the desire to curb the agricultural budget of the EU, will contribute to a further decline of the incomes earned by farmers. In other words, the MacSharry reforms reflect a decline of society's solidarity with the farming sector. The reforms have also broken up the tight and cosy relations that existed between the national Ministries of Agriculture and the farmers' associations. Since 1992, these ministries have been obliged to police farmers much more actively, visiting farms in order to check numbers of cattle and sheep per hectare, amounts of pesticides used, etc. These activities have been much resented by farmers.[62] Also the agricultural reforms have split the agricultural sector itself. In the Netherlands, the Landbouwschap, the professional association that had successfully defended the interests of farmers for many decades, was dissolved in 1997 after much acrimony, and has since been replaced with a more radical organization. Also, in the same country, for the first time the traditional unity between farmers has been broken by the founding of a sectoral union for crop growers.[63] In other words, solidarity with, and between, farmers has been in decline since the early 1990s. Decreasing solidarity stands for a downward movement along the group-dimension. More grid and less group add up to one thing: a movement in the direction of fatalism. The development of more fatalistic social relations within the agricultural sector has gone hand in hand with the espousal of several fatalistic principles/norms, as well as rules/procedures, by the representatives of farmers.[64]

Principles and norms

The fatalist myth of nature is 'nature capricious'. According to fatalists, it is impossible to foresee how ecosystems will evolve in the course of time. No amount of planning or research will be able to influence or predict how natural systems will evolve. This view of nature offers fatalists an additional justification for inaction. The representatives of the agricultural firms in the Rhine countries have seen the Rhine as an extremely resilient ecosystem. They have argued that the fauna and flora of the river cannot have been affected much by the run-off of fertilizers (containing nitrates) and have claimed that farmers themselves often do not understand the governmental concerns with the presence of nitrates in the water as the farmers themselves have not seen any harmful effects from this, viewing the ongoing EU attempts to curb the flow of nitrates into rivers and lakes as exaggerated. At first sight, the view of the resilience of the Rhine defended by farmers' associations seems close to the individualistic view held by large industrial firms in the basin. According to both views, the Rhine river is a resilient ecosystem not much affected by human-made changes. However, there is a crucial difference. The low estimate of the pollution impact of nitrates that has been offered by agricultural organizations contradicts the scientific consensus. This constitutes a huge difference from the view of the Rhine's fragility that has been espoused by large industrial firms who have accepted the scientific evidence of the harmful effects of chemical effluent. Prior to the scientific evidence, they simply had not expected such harmful impacts to exist. As it flaunts scientific consensus, the view of the resilience of the Rhine held by farmers' associations has been more arational, and as such comes closer to a fatalistic view of nature than an individualistic one. The farmers' attitude was nicely illustrated at a symposium organized by the International Commission for the Protection of the Rhine in March 1996. This symposium was the first occasion for all NGOs to discuss their opinions with the ICPR, and was dedicated to the remaining environmental problems in the Rhine basin. After the question of the growing quantities of nitrates had been discussed for some time, a representative of the German farmers rose indignantly to venture: 'Nitrat ist Leben.'[65]

Another regime principle/norm that has been distinguished concerns the trust actors have in each other. Above, it has already been shown that the trust that farmers in the Rhine countries have had in each other, in their representatives and in the Ministries of Agriculture has declined in recent years.

Rules and procedures

In keeping with the above view of the Rhine's resilience, the agricultural sector in the Rhine countries has not espoused a strong sense of responsibility. The representatives of this sector have claimed that farmers have not had the financial means to clean up their effluents. If the governments want to achieve this, they should give farmers extra funding. They have not regarded water protection as an obligation for farmers themselves. In fact, the agricultural associations have tried to turn the tables around, arguing that the existence of farms is an indispensable part of the landscape and environment of water basins. Therefore, the livelihood of farmers should itself be a beneficiary of the environmental protection policies developed by the national and European authorities.[66] These policy views can be labelled 'fatalistic', as they have legitimized inaction on the part of farmers themselves.

When put together, the more fatalistic social environment of farmers, and the fatalistic predispositions that have been expressed by agricultural organizations, easily explain the lack of willingness on the part of farmers to contribute to the restoration of the Rhine. They also make it understandable why farmers have not followed the advice of scientific experts who have found that harvest yields would not diminish if farmers cut their use of fertilizers by half. Fatalists are not risk-takers, and those who move in the direction of fatalism will be less inclined to engage in risky activities.

Conclusion

This chapter has argued that the differences in cooperative behaviour between industrial enterprises, domestic and international governmental organizations and agricultural firms with regard to the environmental protection of the Rhine cannot be fully explained on the basis of conventional factors such as financial resources, pressure from environmental groups or interest maximization. In fact, it was found that only one such factor, varying financial constraints, was helpful in partly answering the puzzle posed in this chapter. It was argued that the various ways of life that Rhine stakeholders have followed provide an additional explanation. Quite divergent actors have populated the Rhine basin, with alternative moralities, lifestyles, perceptions and so on. These factors are important for understanding the different contributions that these actors have made to the restoration of the Rhine catchment area. Existing theories of international relations have often made room for a single rationale among actors but this is surely too limited. The recent history of the clean-up of the Rhine illustrates that there may be a variety of rationales (or ways of life, or cultures) simultaneously at work in international regimes.

This chapter has also provided another example of the contributions that organizations following an individualistic way of life can make to the solution of transboundary issues. In Chapter 4, it was argued that the adoption of individualistic norms and principles advocated by Minister Kroes and the McKinsey consultants revived the intergovernmental efforts to restore the Rhine basin. In the present chapter, it was suggested that the (more) individualistic way of life followed by the large firms in the Rhine area partly explains the extensive investments in water protection that these companies have undertaken. Two elements of this way of life have been especially important: the trust that the large corporations have had in each other, as well as their willingness to take risks.

Overall, the story of the ecological revival of the Rhine has belied the conventional view of environmental politics (existing not only in the academic literature, but also believed in by many practitioners), according to which governmental measures need to force polluters to mend their ways. In particular the large chemical firms have made huge and often voluntary contributions to the environmental protection of the Rhine. Of course, it should not be said that environmental problems do not need much governmental regulation and will always be solved by other forces in society. Clearly, such an anarchistic position is untenable. But perhaps part of the effective governmental regulation of environmental issues lies in stimulating (or at least not thwarting) any contributions that firms may want to make to environmental protection.

Until now, the focus has been on a single case of environmental protection. As a consequence of this choice, institutional factors have not played an important role in the explanations offered above. This is because institutions are treated as background variables, the 'stage' on which the actors play, and a stage which has been more or less the same for all the actors in the Rhine catchment area (at least for the purposes of this study). However, by comparing two cases of environmental protection, one can vary the props and scenery, i.e. the institutional setting in which stakeholders operate, to bring out the impact that political and social institutions have on transboundary environmental issues. The next chapter will compare the environmental protection of the Great Lakes basin (in North America) with that of the Rhine catchment area. This analysis will highlight a number of institutions that tend to diminish the willingness of firms to make voluntary or even legally mandated investments in water protection.

Part III

Institutions Matter As Well

6
Why is the Rhine Cleaner than the Great Lakes?

Introduction

In this book two basic claims are made. The first is that the cultures that inform the thought and behaviour of actors are important determinants of international regimes. In chapters 4 and 5 this is shown for the international regime developed to protect the environment of the river Rhine. It was argued that organizations adhering to the cultures of fatalism, hierarchy, individualism and egalitarianism have made hugely different contributions to the Rhine regime. These organizations have held contrasting perceptions of the issues, have favoured alternative solutions and have been differently affected by the existence of state boundaries. Now the second claim is illustrated, namely that the institutions that structure the relationships between cultures (or, more accurately, between the organizations that adhere to various cultures) matter as well.

This picks up on the work of Frank Hendriks. In a comparative study of the evolving traffic infrastructure of Munich and Birmingham, Hendriks has illustrated the usefulness of combining cultural theory with the set of concerns that have come to be known as the new institutionalism.[1] His general argument holds that one way in which institutions influence policy outcomes is through structuring the relative access that the followers of the four cultures have to the processes of information gathering and searching for solutions as well as making and implementing decisions. In Birmingham since the Second World War the more egalitarian environmental organizations and citizens' groups have continuously been barred from the decision-making processes regarding the city's infrastructure. In contrast, in Munich a much more pluralist regime has existed, in which room has been made for organizations from all cultural backgrounds. Hendriks comes out in favour of the more open policy processes of the city of Munich as they have led to decision-making that may be slower, but that is also characterized by richer and more innovative learning. Michael Thompson has built a similar case for environmental policy processes, arguing that input from all ways of life is a condition for robust measures to protect the environment.[2] In this comparative study the focus will not be so much on institutions that crudely exclude one or more ways of life from political decision-making, but instead will concentrate on the institutions that either divide or unite ways of life. Some institutions, or so it will be argued, bring about a dialogue and mutual understanding between actors, while other institutions tend to sustain antagonisms and miscomprehensions between adherents to different ways of life. These contrasting sets of institutions – through their structuring of the relationships between followers of alternative ways of life – have a significant impact on international issue-areas.

This argument will be illustrated by a comparison between the discharges of toxic substances by the industrial firms bordering the Rhine with the same polluting activities of enterprises situated on the United States side of the Great Lakes. As also explained in Chapter 1, the selection of these cases has been the result of the application of rather strict rules of inference: the rules spelled out by Arend Lijphart's 'comparative cases strategy'.[3] According to this strategy, cases must be chosen: (a) without prior knowledge of the dependent variable; (b) that vary significantly on the independent variable; (c) that are as similar as possible regarding all other potentially explanatory factors. In selecting the cases of the industrial effluents into the Rhine and the Great Lakes these rules have been followed. The dependent variable of this research is the toxicity of industrial discharges into water basins. When this research was started, how the toxicity of the effluents of US firms in the Great Lakes basin would compare to the toxicity of the discharges of the enterprises in the Rhine area was unknown. If anything, the opposite was expected of what was found: during the period from 1970 to the present the industrial discharges from American firms into the Great Lakes basin have remained substantially more toxic than the industrial effluents released into the Rhine. However, there was one significant variation on the independent variable, i.e. the institutional setting. The institutions that regulate industrial emissions from the US side into the Great Lakes divide into two types: (1) those institutions which are derived from the international regime to restore the environment of the Great Lakes, a regime well known for its huge amounts of public participation – something which (as was shown in the previous two chapters) cannot be said for the Rhine regime; (2) other relevant institutions which are part of the relationships between the executive, judiciary, legislature, business community and environmental movement within the United States – again areas that are quite differently organized in Western Europe.

Most other factors that could conceivably provide an explanation for the differences between the cases fall into either of two categories. Various of these potentially explanatory variables have quite similar values across the two cases. During the period of investigation (1970 up to the present), the countries that are represented in both water basins have all been stable democracies, have all experienced high levels of economic prosperity, have shared comparable levels of environmental concern and have all tried to regulate water protection in basically similar ways. Furthermore, the sectoral composition of firms in the Rhine and Great Lakes watersheds have been similar. Both regions have been dominated by steel and iron production, mining, the chemical industry and the production of machinery and equipment. All of these factors can therefore be ignored in the explanation. Furthermore, other elements would lead one to expect the exact opposite of what was concluded. A large number of factors would easily lead one to assume that the industrial effluents from US firms into the Great Lakes would have to be less toxic than the discharges of their European counterparts into the Rhine. For example, in the Great Lakes basin there is a well-organized and influential epistemic community, something which is absent in the Rhine area. Moreover, the international agreements covering the protection of the Great Lakes have also been much more stringent than the Rhine conventions, while the international organization assigned to overview the environmental restoration of the Great Lakes (the International Joint Commission) has had more means to influence policy than its counterpart for the Rhine, the International Commission for the Protection of the Rhine against Pollution (ICPR). Even more importantly, the domestic water protection laws have been stricter in the United States than in any of the Rhine countries.

The puzzle of this chapter can therefore be formulated in the following way: *how is it possible that the discharges by US firms into the Great Lakes have been more polluting than the industrial effluents released into the Rhine, despite the existence in America of stricter national laws and international agreements, more powerful environmental organizations, an influential epistemic community and a more authoritative international organization?* The answer will be formulated in terms of the institutions that have regulated the relationships between the relevant government agencies, legislature, courts, environmental groups and industrial firms, complemented by an explanation of how the role played by the International Joint Commission has amplified some of the adverse effects that US domestic institutions have had on the water protection of the Great Lakes.

The logic of this strategy to compare cases has prevented the choice of a more encompassing dependent variable. In particular, agricultural and municipal discharges of wastewater, airborne depositions of pollutants and the problem of soil contamination have not been considered. Including these factors would not have allowed concentration on institutional factors, as these forms of pollution have been heavily influenced by non-institutional elements such as infrastructure, geography and financial resources. Another methodological admonition has prevented an analysis of the discharges into the Great Lakes from Canadian firms. This omission is perhaps somewhat surprising. The arguments in this chapter will hinge on the differences between American institutions that sustain adversarial relations between governmental and non-governmental actors, and European institutions that tend to bring about more understanding between state and non-state actors. Basically, the chapter will say that the more adversarial processes within the United States have decreased the incentives and motivations of American firms to invest in water protection, whereas the more consensual processes in the Rhine countries have increased the incentives and willingness of industries to clean up their effluents. From this viewpoint it would have been interesting to include in the analysis effluents from Canadian firms released into the Great Lakes basin as Canadian institutions resemble European ones,[4] and their environmental decision-making processes are much more based on consensus than in the United States.[5] The arguments would have been strengthened if the emissions of pollutants by Canadian enterprises in the Great Lakes had been shown to be less toxic than the discharges of US firms. The methodological reason for not doing so has been the need to choose units for comparison that are independent of each other.[6] In the cases of the industrial effluents of the Canadian and US firms around the Great Lakes this condition is certainly not met as firms in the whole area are linked in several ways. A number of US firms own Canadian enterprises in the Great Lakes area. In addition, US firms have part of their production facilities in the Canadian part of the basin. Moreover, commodity trade between Canadian and US firms within the Great Lakes area is extensive. It is no exaggeration to talk of an integrated 'Great Lakes economy'.[7] Canadian and US firms have also combined their efforts to influence policy initiatives in the Council of the Great Lakes Industries (CGLI). As a consequence, it is quite conceivable that the investment decisions concerning water protection that are taken by Canadian firms in the Great Lakes area are influenced by the investment choices of US industries, and therefore indirectly by US institutions. In addition, US environmental policies usually serve as a motor for the development of Canadian environmental policies, making the two cases even more interdependent.[8] To sidestep this problem only discharges by US firms have been considered.

The structure of this chapter closely follows the rationale of the comparative method. First some basic elements of the international regime for the Great Lakes are introduced, as well as the domestic regime to protect the waters of the United States. Thereafter, the puzzle that constitutes this chapter is spelled out. The comparability of the two cases is then considered, followed by an explanation of the puzzle in three steps. First, it is shown that Rhine companies have made extensive voluntary investments in water protection, whereas Great Lakes corporations have grudgingly kept themselves to a bare legal minimum. Subsequently, it is shown that the reluctance of Great Lakes firms to invest in anti-pollution measures is part of a broader set of antagonisms that have greatly divided the organizations involved in the Great Lakes watershed – as opposed to the Rhine valley. Thereafter, the institutional factors at both the domestic and international level are highlighted that have fuelled these divisions: (a) American exceptionalism; (b) the institutions that shape the relationships between public and private organizations in the involved countries; and (c) the international regimes for the protection of the Great Lakes and the Rhine. After this institutional explanation has been presented the strength of an alternative, more parsimonious account based on rational choice-theory is considered. Finally the extent to which these findings can be generalized to other cases of environmental protection is taken into account.

A final note on how the data for this chapter were gathered. In addition to the 59 interviews with stakeholders in the Rhine basin that were used in previous chapters, 48 interviews were held with representatives from industry, federal and regional government, international organizations and environmental groups in the Great Lakes basin and Washington, DC. (see Appendix B). The information obtained during these interviews was processed in the informal way described in Chapter 1.

Scene-setting: the Great Lakes[9]

As is the case with the Rhine river, the Great Lakes in North America form a site where important ecological values and huge economic interests come together – which make them an interesting topic for political science research. Each of the five Lakes (from west to east: Lake Superior, Lake Michigan, Lake Huron, Lake Erie and Lake Ontario) is massive. The Great Lakes basin contains diverse ecosystems. The northern parts of the basin are very cold (with daily averages in January of more than 10 degrees Celsius below zero) as they receive Arctic winds. The shores in these parts are rocky and acidic, and coniferous forests dominate the landscape. The southern areas of the Great Lakes basin are warmer, while the soils are fertile as they consist of mixtures of clays, silts, sands, gravels and boulders. The remaining forests are deciduous. The Great Lakes basin contains unique wetlands such as the Long Point complex and Point Pelee on the northern shore of Lake Erie, as well as the National Wildlife Area on Lake St Clair. The Lake Ontario basin includes Niagara Falls. Before the European settlement of the area, there were an estimated 180 fish species indigenous to the Great Lakes, including herring, blue pike, lake trout, walleye, deepwater cisco and lake whitefish. On the shorelines, foxes, wolves, beavers and muskrats roamed. In the air, there were cormorants, herring gulls and eagles. A number of these species remain today or have returned, but others have died out or emigrated from the basin as a result of human activities.

The human activities in the Great Lakes basin that have led to declining numbers of animal species and other forms of environmental degradation are numerous. Municipalities and firms in the basin have discharged polluted wastewater into the lakes and connecting rivers. Outside the region, industries have released pollutants into the air that have ended up in the Great Lakes. The pesticides used on farms have reached the lakes through groundwater. Urban sprawl and the cultivation of land have destroyed habitat while over-fishing has severely depleted fish stocks. One major environmental problem is only indirectly related to human activities: the appearance of exotic species into the lakes. These are animal species coming from other ecosystems, whose arrival upsets the balance within an ecosystem. In particular the arrival of sea lamprey and the introduction of the zebra mussel has created many problems for indigenous species in the Great Lakes.

The economic importance of the Great Lakes area is considerable. Of the Fortune 500 largest industrial companies in the United States, almost half have their company headquarters within the Great Lakes states. Of the 500 largest non-industrial companies in the United States, about two-fifths are in these States.[10] Around 33 million people live in the Great Lakes basin, a large number of them in the metropolises of Chicago, Detroit, Toronto and Montreal. About 47 per cent of this total population draws its drinking water from the Great Lakes. Shipping on the lakes (especially of iron ore, coal and grain) totalled 60 million tonnes in 1990. Sports and food fishing generated revenues of more than $6 billion in 1985 (the latest year for which binational data are available).[11] Manufacturing in the Great Lakes states comes to a larger percentage of total output than in the rest of America, at present hovering around 20 per cent of total output. Over 4 million people have been employed in the manufacturing industry in the Great Lakes states during recent years.

This confluence of ecological vulnerabilities and economic opportunities has given rise to a host of political arrangements to manage the Great Lakes ecosystems. Many political entities are involved. The US–Canadian border runs through the middle of four of the five lakes. Only Lake Michigan is completely within the United States. The basin envelops two Canadian provinces (Ontario and Quebec – the latter as the St Lawrence river runs through it), as well as eight US states (Minnesota, Wisconsin, Illinois, Ohio, Indiana, Michigan, Pennsylvania and New York). This multitude of political territories has necessitated the development of organizations and institutions for the protection of the Lakes at all levels: international, national, regional (inter-state) and state.

The major international organization dealing with environmental concerns in the area is the International Joint Commission (IJC). Created in 1909 by the Boundary Waters Treaty between the federal governments of Canada and the United States, it was expected to assist the governments in solving transboundary problems, including environmental issues. Although the Commission's area of attention is the whole of the US–Canadian border and includes more than environmental matters, the bulk of its activities has been related to the protection of the Great Lakes. The IJC is led by six Commissioners, half of whom are nominated by the President of the United States, and half on the advice of the Canadian Prime Minister. The Commissioners can rely on a permanent staff. In addition, the IJC has set up more than 20 boards, which have been filled by officials from the states, provinces and federal governments involved, as well as by scientific experts and people working for citizens' groups. Two influential such boards on pollution matters have been the Water Quality Board and the Scientific Advisory Board.

In part, the IJC works on the basis of references. This means that the Commission advises the two federal governments on how to solve transboundary issues whenever it is asked to do so by those governments. It also has binding conflict-resolution powers. Any plans for the construction of dams and hydroelectric power that might affect water levels upstream and downstream can only be implemented after the Commission has given its consent. Last, the governments have asked the IJC to monitor, and regularly comment upon, the implementation of various environmental treaties between Canada and the United States. As a consequence, the International Joint Commission has long been a major actor shaping the international regime for the restoration of the Great Lakes.

Several official agreements are part of this international regime. The 1972 Great Lakes Water Quality Agreement was aimed at reducing eutrophication, as well as visible pollution. Its successor, the 1978 Great Lakes Water Quality Agreement was more ambitious and called for the 'virtual elimination' of persistent toxic substances and the restoration of the integrity of the Great Lakes ecosystem. The 1987 Protocol to the latter treaty promised the development of 'Remedial Action Plans' (RAPs) in what are described as 43 'Areas of Concern'. The 1987 Protocol also establishes a need for the development of 'Lakewide Management Plans' that propose actions to improve the quality of the water in each individual lake. Both the RAPs and the Lakewide Plans are currently still under construction.

Non-governmental transboundary associations have also been active in the Great Lakes watershed. The first binational environmental group, Great Lakes Tomorrow, functioned in the 1970s and 1980s.

In 1982 Great Lakes United (GLU) was formed as an umbrella organization of more than 200 environmental groups from around the basin. GLU has been an influential actor in the international regime. Only since the beginning of the 1990s has industry attempted to directly influence international policy developments. In 1991, the Council of Great Lakes Industries was formed to represent the common interests of United States and Canadian corporations from the manufacturing, utilities, transportation, communications and financial services and sectors that have investments in the Great Lakes basin.

Other governmental actors in the basin that have dealt with environmental issues are a cross between inter-state and international organizations. Both the Great Lakes Commission and the Council for the Great Lakes Governors are primarily inter-state bodies, but cooperate with the two Canadian provinces involved (Ontario and Quebec). The Great Lakes Commission lobbies the federal governments on behalf of the states and advises state agencies and provincial ministries on policies for environmental protection and economic development. It also acts as a clearing-house for technological, legal and other information that may be of value to organizations within the Great Lakes area. The Council of the Great Lakes Governors is more sophisticated, providing a venue for the Governors to coordinate major basin-wide policy initiatives regarding economic development and ecosystem management. The First Nations (which live primarily in Minnesota, Wisconsin, and Michigan) are represented by the Great Lakes Indian Fish and Wildlife Commission, which has become quite vocal in recent years.

Besides all of these basin-related treaties and organizations, the Great Lakes are also affected by domestic water protection laws and policies as well as by national organizations. Since its inception in 1970, the US Environmental Protection Agency (EPA) has been responsible for the protection of American waters against pollution. Other federal departments and agencies have of course also been involved (including the Fish and Wildlife Service, the Army Corps of Engineers, the Department of Agriculture, the Department of the Interior and the Department of Commerce – the latter especially as it houses the Great Lakes Environmental Research Laboratory). But their activities have focused more on research and the restoration of habitat. The EPA has been the lead agency for the environmental protection of water bodies. The system that has been set up by the EPA with regard to industrial discharges into open waters can be described as follows. Industries need a licence from the EPA to discharge effluents into rivers and lakes and this permit stipulates the conditions that its emissions have to meet calculated on the basis of two sets of criteria. The first consists of effluent limits, which spell out the maximum amount of pollutants per unit of water that an industrial emission can contain. Such limits exist nation-wide for a number of pollutants. The second set of criteria pertains to water quality standards which determine the minimally required biological and chemical conditions of surface waters, for example the minimum acceptable level for dissolved oxygen in a river or lake. If, in a particular case, these water quality standards are not met with the help of the 'normal' effluent limits, the EPA has the right to impose stricter conditions on industrial emissions. The EPA usually leaves the actual process of granting permits to the relevant state agencies. The states must at least meet the standards laid down by the EPA but can also impose stricter limits. Minnesota, for in-stance, is a Great Lakes state that has sometimes followed stricter rules. If the EPA judges that states have not adhered to the mini-mum standards, it will take the process of granting permits into its own hands, which is what happened to Pennsylvania in the 1990s. Besides imposing effluent limits, the EPA often also prescribes the technological means which enterprises have to use in their efforts to meet the standards. In other words, the EPA has traditionally followed a strict 'command-and-control' approach.

The EPA and the other federal agencies and departments involved base their authority on a variety of national laws. The most impor-tant of these (for the clean-up of the Great Lakes) is the Clean Water Act adopted in 1972 and amended several times since. On one occasion federal legislation was adopted that pertained only to the Great Lakes ecosystem: the 1990 Great Lakes Critical Programs Act. Under the terms of this Act the EPA issued the Water Quality Guidance for the Great Lakes System in 1995. This Guidance has created uniform water quality standards for the US part of the whole basin.

Nationally organized environmental organizations and business associations have also tried to influence domestic water policies in general and Great Lakes initiatives in particular. In the green corner can be found: the World Wide Fund for Nature (WWF), the Sierra Club, Nature Conservancy, the League of Women Voters, Greenpeace and the Environmental Defence Fund, among others. These organi-zations have participated in the umbrella organization Great Lakes

United, but have also undertaken many activities on their own. In the ring for business have been: the Chlorine Chemistry Council, the American Forest and Paper Association, the Chemical Manufacturers Association, the National Mining Association, the American Iron and Steel Association and the Great Lakes Water Quality Coalition. Individual firms also lobby quite often for themselves.

A puzzle

During recent years, the emissions of US corporations into the Great Lakes have contained a number of persistent toxic substances that have already been kept out of the effluents of enterprises in the Rhine valley for quite some time. This has occurred despite the existence of many differences between the two cases that would have led one to predict the opposite. These include:

• an influential and large international organization in the Great Lakes area (the International Joint Commission), as compared to the weak and small International Commission for the Protection of the Rhine against Pollution (ICPR);
• environmental organizations in the Great Lakes region that have been better organized and have had more access to political decision-making than similar groups concerned with the Rhine;
• a thriving and well-organized epistemic community pushing for the restoration of the Great Lakes – something virtually absent in the Rhine basin;
• international environmental treaties in North America that have been stricter and more influential than the international agreements on the protection of the Rhine;
• excellent, decades-long international cooperation between Canada and the United States on the protection of the Great Lakes, and (until 1987) internecine diplomatic strife in the Rhine basin;
• domestic water protection laws in the United States that have been more stringent than anywhere in Europe.

All of these assertions will have to be substantiated. First, it will be shown that the effluents of US firms discharged into the Great Lakes have remained more toxic than the emissions of corporations discharged into the Rhine.

Effluents compared

Comparing the toxicity of effluents entails deciding which substances should be seen as pollutants. This is difficult, since part of the struggle between government agencies, firms and environmental groups is exactly about which substances are, and which are not, toxic (and about the concentration levels at which substances are, or are not, harmful). Given these disputes, we have opted for a minimal set: a list of persistent chemical substances that most involved organizations in both regions have accepted as highly dangerous. In 1991 the IJC listed 11 'critical pollutants'. This list contains chemicals that have widely been viewed as highly toxic in both the Rhine and Great Lakes basins. According to the IJC and the US Environmental Protection Agency, these 11 pollutants are still being released into the Great Lakes by US firms.[12] Five pollutants are not relevant here, as they are solely discharged by farms. The question now becomes: are the remaining six pollutants still flushed into the Rhine?

An answer to this question can be construed on the basis of the fine-grained measurement systems operated by the water supply companies on the Rhine (organized into the IAWR) as well as by the ICPR. These two monitoring systems are so elaborate as to make it possible to deduce the toxicity of industrial discharges into the Rhine from the river's water quality. (It is not possible for chemicals that have been released by Rhine companies to escape registration before they flush into the North Sea.) For this analysis, 1996 data were relied on, regarding water quality in a downstream part of the river, namely in the Dutch town of Lobith. From this, Table 6.1 can be constructed.

Table 6.1 shows that in 1996 three of the six persistent toxic substances that were still being discharged by Great Lakes firms could not be detected in the Rhine. This means that these pollutants were no longer present in the effluents of Rhine corporations. Three other substances (PCBs, mercury and lead) did show up in the Rhine water. The reason that PCBs were still found in the Rhine water has to do with previous discharges of these substances. In 1996, PCBs were not dumped into the Rhine, but in the 1970s and 1980s they were. These substances then sank and polluted the river's sediment, and due to currents in the river, PCBs are sometimes released from the riverbed which explains why PCBs can still be found in the waters of the Rhine.[13]

The other priority substances that were still detected in the Rhine are mercury and lead. These two substances are quite different from the others in Table 6.1. Mercury and lead are mainly formed in natural processes, while the others are human-made so it is not realistic to demand zero levels of mercury and lead. For these sub-

Table 6.1 Quality of Rhine water at Lobith in 1996 (annual averages in μkg/l; the symbol '<' stands for quantities too small to be detectable with current measurement techniques)

	Rhine water in 1996
PCBs*	6–12
Dioxin (2,3,7,8-TCDD)	<
Furan (2,3,7,8-TCDF)*	<
Mercury	0.038
Benzo(a)Pyrene	<
Lead	4,5

* Figure received from the ICPR. Other numbers taken from IAWR 1996.

stances another comparison is needed. Luckily, the Rhine water supply companies have established stringent standards for the water quality of the river. If these standards are met, then it would be possible to prepare safe drinking water with natural purification methods. The levels of mercury and lead in the Rhine were well within these strict standards.[14]

Perhaps, prima facie, these results do not seem impressive. However, these six substances are among the most dangerous and persistent chemicals used by modern industry. Moreover, these chemicals stand for large categories of pollutants. PCBs come in 209 varieties, dioxin is a family of 75 chemicals and furan has 135 variations. The finding that these six highly toxic substances are no longer emitted into the Rhine indeed represents a major difference between the two cases. The finding corresponds with expert opinion. During the interviews with stakeholders from the Great Lakes watershed a large majority said that corporations still emitted small quantities of traditional pollutants such as PCBs. In the Rhine basin, the predominant opinion among the interviewees was that firms no longer released such pollutants. This corroborates the finding of Christopher Allen that the German chemical enterprises invested twice as much in environmental protection as their American counterparts from the mid-1970s to mid-1980s.[15] It can therefore be concluded that Rhine companies have purified their releases to a greater extent than Great Lakes corporations. This conclusion is striking given the many factors that would have induced one to expect the opposite. These factors are considered below.

Two international commissions

The International Joint Commission (IJC) has had more means for promoting environmental protection measures in the Great Lakes basin than has the International Commission for the Protection of the Rhine (ICPR). First, the powers and responsibilities of the IJC have been more comprehensive than those of the ICPR. For instance, the IJC has veto rights over any project that affects lake levels, and can be requested to exercise binding arbitration powers in transboundary disputes. Moreover, the IJC has had a vastly larger budget (on average about eight times that of the ICPR) and staff.

The influence of environmental organizations

In addition, the environmental organizations concerned with the Great Lakes have been better organized and more influential than the ecological groups in the Rhine valley. With the formation of Great Lakes United in 1982, the former environmental groups reached a unique degree of cooperation. Through Great Lakes United, environmental groups can speak unequivocally to decision-makers. Such concerted effort between citizens' organizations has been fully absent from the Rhine basin.

Furthermore, Great Lakes environmental associations have had superior access to (inter)governmental decision-making. Under the Administrative Procedure Act since 1946, it is obligatory for US agencies to seek public input whenever they contemplate new policies and laws. Therefore, US agencies cannot develop new water protection policies without extensive public hearings and also involving environmental groups. In none of the Rhine countries has the participation of environmental organizations in domestic decision-making been so firmly ensconced in domestic law. Internationally, the IJC has offered Great Lakes environmental movements unique opportunities and has provided them with several platforms. The IJC organizes Biennial Conferences at which environmental groups comment on the progress made by the North American governments under the 1978 Great Lakes Water Quality Agreement. The IJC then reports these comments to the governments and media. Moreover, environmental activists have sat on many advisory boards to the IJC. Even more remarkable is the role that three environmental organizations played during the international negotiations over the 1987 Amendments to the 1978 Great Lakes Water Quality Agreement. During these negotiations, Great Lakes United, the Sierra Club and the National Wildlife Federation participated as observers on the US negotiation team. The Canadian diplomatic mission included two members of Great Lakes United as observers.[16] In contrast, the ICPR only started to communicate with environmental groups in 1996.

Last, American environmental organizations have unusual standing in courts of law. When Congress adopted the Clean Water Act in the early 1970s, it allowed environmental organizations and other interest groups to sue both polluters and the EPA if the law's aims were not being achieved. Thus, environmental groups can push the EPA into implementing policies that meet the standards of the Clean Water Act. As will be shown later, Great Lakes environmental organizations have often used these legal possibilities. This is again in contrast to Western Europe, where the possibilities for NGOs to take legal action have always been smaller. In sum, environmental groups have had much more scope to influence policies in the Great Lakes basin than in the Rhine region.

One epistemic community

The term 'epistemic community'[17] seems to have been invented for application to the case of the protection of the Great Lakes. An extensive and well-organized epistemic community of law professors and natural and political scientists has advocated the protection of the Lakes since the 1970s. Two journals are devoted to publishing their research: the *Journal of Great Lakes Research* and the *Toledo Journal of Great Lakes' Law, Science, and Policy*. Over the past 25 years, four series of meetings have been convened under the title 'Canada–United States Inter-University Seminars for the Great Lakes', bringing together members of universities, government agencies and environmental organizations from around the basin to discuss Great Lakes protection issues.[18] In the Rhine valley, no traces of any such epistemic community were found.

Together, the IJC, environmental organizations and the epistemic community have formed a formidable force for the protection of the Great Lakes. Many cross-cutting linkages exist between the three groups of actors.[19] In the Rhine region no such impressive array of 'green' forces has ever seen the light of day.

How strict have international treaties been?

The international agreements for the protection of the Great Lakes have been more ambitious than those concerning the Rhine. As early as 1972 the first Great Lakes Water Quality Agreement was signed, aiming at the reduction of phosphorus. This was four years before the Rhine Salt and Chemical Conventions were signed. In 1978 a second, much more comprehensive Great Lakes Water Quality Agreement came into force, aiming at restoring 'the chemical, physical, and biological integrity of the Great Lakes Basin Ecosystem', and calling for the virtual elimination from the Lakes of 'any or all persistent toxic substances'. The wording of the 1976 Rhine Chemical and Salt Conventions is much more cautious. For instance, the eco-system concept does not appear in these treaties. The concept only appears in the Rhine regime with the adoption of the Rhine Action Programme in 1987. Furthermore, the Rhine Conventions from 1976 merely express the intention to eliminate emissions of a limited black list of substances, and to reduce emissions of an equally limited grey list of pollutants.

Amicable versus internecine international relations

The strict international agreements concerning the Great Lakes sprung from almost frictionless negotiations between the governments of the United States and Canada. Since the early 1970s, these negotiations have taken place in an atmosphere of great cooperation. Again the Rhine regime forms a striking contrast, as was shown in Chapters 4 and 5 which described how ineffective and conflictual international cooperation on the Rhine was – at least before 1987. It was only after the Sandoz incident in late 1986 that the Rhine states started to implement some effective international measures (the Rhine Action Programme). However, as argued in Chapter 5, the Rhine Action Programme did not have a great influence on the ongoing efforts to purify industrial discharges, but did much more to stimulate habitat restoration in the basin.

The strictness of domestic laws and policies compared

Another element of the puzzle of this chapter concerns domestic water protection laws. The domestic laws pertaining to the Great Lakes appear to have been stricter than those relevant for the Rhine. Again, this would lead one to erroneously assume that the effluents of Great Lakes firms had been less polluting than the discharges of Rhine firms.

The water protection programmes of all involved countries have had the same basic features.[20] In all countries, a firm can only discharge wastewater if it has a permit to do so. This permit lists the conditions that industrial discharges have to meet, based on effluent limits and water quality standards.[21] It would have been most appropriate to have compared European and American effluent limits but unfortunately this is not feasible. A first hurdle is that countries have regulated different pollutants. Effluent limits pertain to categories of chemical substances and the involved governments have categorized substances differently, which makes cross-national comparison difficult. Moreover, in most countries specific effluent guidelines are usually developed for each particular branch of industry. However, the countries also categorize industries differently, making a comparison of effluent limits even harder.

Yet, something can still be said about the relative stringency of the domestic water protection programmes, starting with the aims of water protection laws. The US Clean Water Act has included more ambitious goals than the European laws. In 1972, this Act required the nation's waters to become 'fishable and swimmable' by 1983. They also established a national goal of 'zero discharge' of pollution by 1985. Given the highly polluted nature of the American waterbasins in the early 1970s, these were ambitious plans indeed. In addition, the EPA was required to develop effluent standards regardless of their technological or economic achievability.[22] The aims of the European water protection laws have remained much more modest. Moreover, these documents invariably stipulate that water protection should be balanced against the protection of financial interests.

The argument continues with technology-based standards. Under the US Clean Water Act, dischargers were obliged to employ the 'best practicable technology currently available' by 1 July 1977, and the 'best available technology' after 1 July 1984. These standards have been much more binding than the European ones. They call, respectively, for the installation of wastewater plants with an 'average of the best' and 'best of the best' performance within an industry. The contrast with Europe is huge. For example, before 1990 German firms only needed to apply 'generally accepted technology', which did not really impose anything on industry. Since 1990 German firms have been obliged to use the current 'state of technology', which again is a rather lax standard, and similarly relaxed policies have existed in the other Rhine countries. Given

Table 6.2 An international comparison of water quality standards (in µg/l; bold numbers express the stricter standards)

	Great Lakes (US side)	Rhine
Arsenic	**147.9**	40 000
Chromium	**10.98**	100 000
Mercury	**0.0018**	500
Dieldrin	**0.00041**	0.001
Hexachlorobenzene	**0.00045**	0.001
Endrin	0.036	**0.001**
Lindane	0.5	**0.002**
Benzene	310	**2**
DDT	**0.00015**	0.001
PCBs	**0.0000039**	0.0001
Trichloroethylene	370	**1.0**

Sources: 40 CFR, ch. 1 (7 January 1999), 640–709; ICPR, 1993.

the huge variation between American and European technology standards, it can safely be assumed that the effluent limits in the United States have been more severe than in the Rhine countries. This is because effluent standards are calculated with one eye on technology standards.

A last issue usable for comparing the strictness of domestic laws is the difference between water quality standards. In 1995 the EPA issued the Water Quality Guidance for the Great Lakes System. This document harmonizes water quality standards across the Great Lakes states. In 1991, a common set of water quality aims for the Rhine was also adopted by the involved governments. These basin-wide standards were more stringent than anything that had existed before in the individual Rhine countries. The two lists can partly be compared.

Table 6.2 shows that for seven out of 11 pollutants, water quality standards for the Great Lakes are presently stricter than those for the Rhine basin. This finding is merely suggestive. It does not warrant the conclusion that water quality standards pertaining to the Great Lakes have always been more stringent, because it only compares a limited number of current standards. Still, it is a further piece of evidence pointing to the conclusion that the domestic water protection laws covering the Great Lakes have been stricter than the national laws concerning the Rhine. This finding ties in with the more ambitious legislative aims and more stringent technological standards that have existed in the United States. On the basis of all these factors, it seems safe to assert that the domestic water laws impinging on the (US side of the) Great Lakes have been stricter than those pertaining to the Rhine.

All the differences between the two cases described above point to just one conclusion: the industrial discharges into the Great Lakes must have been less toxic than the emissions of firms into the Rhine. Still, the opposite has occurred. Before an institutional explanation of this puzzle is offered the utility of comparing these two cases of water protection will be considered.

The comparability of the two cases

Instead of looking for an answer to the puzzle of this chapter, one could question the validity of any comparison at all between the environmental protection of the Rhine and the Great Lakes. The latter are after all immense lakes, while the former is not even Europe's largest river. It could be argued that it does not make sense to compare two watersheds that are so dissimilar in their natural conditions. The first way in which the comparability of the two cases has been ensured is by limiting the dependent variable of the research to industrial effluents. As such, the dependent variable includes only human-made effects on the watersheds – and is not influenced by any natural differences between the basins.[23] Another way of ensuring the comparability of the two cases is by treating the natural differences between the watersheds as potential explanatory variables. The possible explanatory power of these independent variables must be considered. It could especially be speculated that the vastness of the Great Lakes has swayed industrialists. Surely a bit of pollution from a Great Lakes firm cannot do much harm to such immense waters? Such reasoning might have played a role. However, a counterargument also exists. The Great Lakes are not only enormous bodies of water; unsurprisingly they are also lakes, whereas the Rhine is a river. As a consequence, substances that enter the Lakes stay there for a much longer period – 191 years for a raindrop falling into Lake Superior as compared to at most 77 days for the Rhine. Therefore, pollutants that enter the Great Lakes bioaccumulate much more. During bioaccumulation, the concentration of chemical substances within living organisms increases manyfold with every next link in the food chain. The predators on top of the food chain (eagles, salmon) therefore accumulate a level of pollution in their bodies that is several thousand times higher than the pollution level of the water. Many people in the Great Lakes basin have been aware of this process, since it has led to a ban on the sale of various fish from the Great Lakes for several decades. The problem is also invariably mentioned in informative brochures on the Lakes. So, the fact that the Great Lakes are massive lakes, whereas the Rhine is a river, cuts both ways. Perhaps the vastness of the Great Lakes has induced business leaders to discount the impact of pollution emanating from their firms. But it is equally plausible that Great Lakes industrialists have been aware of certain environmental problems (such as bioaccumulation and sediment pollution) that are far greater in lakes than in rivers. It is therefore quite difficult to argue that the natural differences between the ecosystems have had a strong, unequivocal influence on the dependent variable of this research.

The comparability of the two cases could also be challenged differently. The puzzle could be 'wished away' by observing that the Rhine is an exceptionally successful case of environmental protection. As shown in Chapter 4, the Rhine regime has come to be seen as an example for other European attempts to restore aquatic ecosystems. Perhaps the puzzle is merely a coincidence caused by pairing a successful European effort with an ineffective American attempt. What this argument leaves out is that the protection of the Great Lakes is widely regarded as one of the most successful environmental efforts in the United States. It has been noted that the Great Lakes have been cleaned up to a larger extent than other US basins.[24] Moreover, the international regime for the Great Lakes has frequently been held up as an example for other countries.[25] The protection of the Lakes has been as much a 'role model' in America as the clean-up of the Rhine has been in Europe. This makes it all the more interesting to compare the two cases.

Voluntary steps versus feet-dragging

The finding that the industrial discharges into the Great Lakes have remained more toxic than the industrial effluents discharged into the Rhine, despite stricter domestic legislation in the United States, logically entails one, or both, of two things. US water laws must have been badly implemented, and/or voluntary investments in water protection (i.e. investments not legally required) must have been more extensive in the Rhine valley. In reality, both developments have taken place simultaneously.

The Rhine companies have taken many voluntary measures to clean up their effluents. This was shown in detail in the previous chapter, which argued that by the mid-1980s the Rhine firms had greatly and voluntarily reduced their discharges into the river. Almost the opposite has taken place in the Great Lakes area. The effluents of US firms into the Great Lakes have been regulated by the 1972 Clean Water Act. David Vogel has documented the slow implementation of this Act.[26] From the start, the EPA was overwhelmed by the challenges involved in implementing the Clean Water Act. The law called upon the EPA to develop and scientifically justify many detailed quality standards and effluent guidelines but the agency's budget was insufficient to meet these tasks on time. Nor did regulated firms accept the EPA's standards and reacted by lobbying Congress and the White House and by suing the agency. One EPA interviewee estimated that about 90 per cent of all water regulations developed since the 1970s has been challenged in court, and the court cases have sometimes taken years to be settled. And even when its water regulations were upheld, the EPA sometimes had to force firms and states into compliance by strict monitoring and legal action. All these processes have slowed down the implementation of the Clean Water Act. In 1989 the EPA conceded that over two-thirds of the nation's wastewater treatment plants failed to comply with the Clean Water Act.[27]

It is safe to assume that a number of these inadequate facilities were also located in the Great Lakes watershed. This is the case, since Great Lakes corporations have been at the forefront of industry's resistance against the water protection efforts of the EPA. Great Lakes executives have often viewed the stipulations of the Clean Water Act as too strict, unfair and cost-ineffective, and have regarded the EPA's attempts to implement the Act as too heavy-handed.[28] Moreover, Great Lakes executives have acted on their beliefs, by taking a leading role in the fight against the Clean Water Act. Both Great Lakes firms themselves and their interest associations (such as the American Automobile Manufacturers Association, which is dominated by the Michigan carmakers) have battled at various fronts to get the Clean Water Act watered down: in the courts, before Congress and at state level. References to these struggles will follow. Suffice it here to conclude that as Great Lakes enterprises have waged prolonged battles to thwart implementation of the

Clean Water Act, it is unlikely that they have taken more water protection measures than the legal minimum necessitated. These contrasting developments form an answer to this chapter's puzzle: *in the Rhine basin industrial effluents have been less toxic than in the Great Lakes watershed because Rhine firms have made extensive voluntary investments in water protection, while Great Lakes corporations have kept their protection efforts limited and have tried to reverse US water laws.*

These different inclinations of Rhine and Great Lakes executives to invest in water protection have been rooted in larger processes. The politics of the Great Lakes protection have in general been more adversarial than the politics of the Rhine restoration. In the (US side of the) Great Lakes region, firms, environmental groups and government agencies have disagreed more vehemently about a greater number of issues. They have not only quarrelled about the solutions to the ecological problems, but have also remained bitterly divided about the gravity and sources of these problems. Of course, the politics of the Rhine protection have not been wonderfully harmonious – many disagreements have indeed existed with regard to the causes, extent and solutions to the Rhine's ecological problems. However, compared to the Great Lakes, these disagreements have been more contained and less bitter. The comparatively antagonistic processes in the Great Lakes region have been intertwined with the lesser willingness of Great Lakes companies to invest in water protection. Doubting the seriousness of the pollution of the Lakes, and disagreeing with the measures proposed by the EPA and environmental organizations, Great Lakes enterprises have frequently decided to keep their anti-pollution measures limited – at least when compared to Rhine corporations. This is what initially explains the puzzle of this chapter.

The above arguments can also be expressed in the language of cultural theory, though this is not strictly necessary for the sake of this chapter. However, it is essential for the overall message of this book (namely that both cultures and institutions matter). Furthermore, it allows for the construction of a richer picture of the motivations and actions of the organizations involved. In the terms of cultural theory, American institutions tend to polarize the opinions and behaviour of (members of) organizations that are characterized by alternative political cultures. Below will first be described the ways of life adhered to by the various actors in the Great Lakes basin. Then it will be argued which opinions and perceptions in particular have been polarized in the Great Lakes watershed.

The cultural configuration of the stakeholders in the Great Lakes ecosystem is somewhat comparable to that of the Rhine actors.[29] As is the case in the Rhine regime, business corporations tend to display a more individualistic behaviour and mindset. This starts with their optimistic view of the resilience of ecosystems. In their view, the Great Lakes watershed has already been cleaned up to a significant degree. They acknowledge that there are still some local problems, but overall they rate the health of the Great Lakes ecosystem as quite satisfactory. Business representatives tend to underplay the toxicity of some chemicals that other groups of actors regard as poisonous. Connected to this is their opposition to the precautionary principle. Often they only want to accept the existence of an environmental problem when conclusive and abundant scientific evidence has been gathered. In addition, business managers are very concerned with cost-efficiency. Environmental measures should be taken where the biggest environmental benefits can be got for the lowest costs. The entrepreneurs point out that at present reductions of agricultural discharges would result in higher marginal benefits for the environment than any further cuts in industrial emissions. They also favour simple, cost-effective solutions. For instance, in at least 42 places in the Great Lakes basin, sediments are currently heavily polluted with chemicals and heavy metals. Government agencies and environmental groups insist that these places should be restored to a pristine state by removing all pollutants from the soil. Business spokespersons are against this, on the grounds that during dredging many chemicals would become detached from the soil and would thereby be re-released into the Great Lakes ecosystem. Their favoured solution for a number of places is to leave the polluted sediments untouched (or to cover them up). Business representatives also show a willingness to engage in cost–benefit calculations of pollution. They sometimes argue that the vices of pollution should be weighed against the virtues. On the one hand, pollution causes environmental degradation and, occasionally, health problems. Yet, on the other hand, it is also a by-product of economic growth and activity. Sometimes these virtues simply outweigh the vices. To business executives, 'zero discharge' of toxics (the ultimate aim of the Clean Water Act and the Great Lakes Water Quality Agreement) simply does not make sense. However, such cost–benefit reasoning is usually anathema to environmental groups.

Business people also tend to oppose the prescription of 'best available technology' – not only because technological standards are quite strict in the United States, but also on principle. They argue that an industry-wide prescription of technology is often not the most efficient, or even most effective, way of reducing pollution in specific plants. They feel that decisions concerning production processes are properly left to the corporations themselves instead of the government. In general, business leaders feel that environmental governance should be limited and local. For instance, some of them have opposed the harmonization of water quality standards in the whole basin under the 1990 Great Lakes Critical Programs Act. In their view, the natural conditions and pollution problems of each lake are quite different from those of the others. Therefore, allowance should be made for local variations of environmental standards. Business executives also refrained from engagement with the International Joint Commission, at least until the early 1990s, when they seriously started to debate among themselves whether to lobby for the abolition of this 90 year-old international organization. In the end they decided not to do so, but the discussion partly reflects the dislike among corporate leaders of high-level government. In conclusion, the behaviour and mindset of business leaders show clear signs of the individualistic culture. Their view of the environment, preferred environmental policies and approach to governance have all been individualistic.

Also comparable to the Rhine case is the cultural disposition of the involved environmental organizations. As a whole, the environmental groups concerned with the protection of the Great Lakes tend to have more egalitarian attitudes than most other stakeholders in the basin. They share their more egalitarian views with the epistemic community of academics. This is above all evident in the mode of governance favoured by most environmental organizations and concerned academics. Usually, they strongly prefer public participation in decision-making at all possible levels. With 'public participation' they mean involvement of 'ordinary citizens', and not business executives. Members of environmental groups see themselves as representatives of ordinary American citizens. The public participation programmes of the International Joint Commission come close to their ideal of international governance. Another egalitarian trait of the environmental groups and academic community is their distrust of industry and government agencies. In their view, industry will only reduce pollution when forced to do so by others. Business leaders are often thought to have little regard for environmental protection. The EPA is sometimes exempted from a general suspicion of the motives and capacities of governmental actors, but state agencies and (especially Republican) politicians are usually viewed with distrust by environmental movements.

Their idea of the robustness of the Great Lakes ecosystem tends to be almost the opposite of the view cherished by the business community. According to spokespersons of environmental groups and academic writers, the Great Lakes form a highly fragile and sensitive ecosystem. All polluted sediments should therefore be fully restored. Zero discharges of persistent toxics should be attained as soon as possible. As this has not yet taken place, the Great Lakes are sometimes judged to be in a worse shape than ever – an evaluation that is in sharp contrast to the opinions expressed by business executives. Almost needless to add, environmental groups are firm believers in the application of the precautionary principle to a wide range of chemicals. They tend to view any chemical substance as potentially toxic, and environmental activists frequently speculate about the hidden, incremental dangers that chemical substances can pose. At the time of writing, a huge controversy has taken place in the United States over a book co-written by Theo Colborn, a zoologist with the World Wildlife Fund who has also written extensively on Great Lakes matters.[30] The main thesis of this book was that our 'fertility, intelligence and survival' is threatened by the spread of industrial pollutants through the web of life (mainly through the alleged effects of these pollutants on sexual development and reproduction). The scientific evidence on which the book is based was dismissed out of hand by the business representatives interviewed, although it has received support in many other quarters.

One element is conspicuously absent from the egalitarian discourse of environmental movements and academic community: a fundamental critique of capitalism and the profit motive. In the Rhine valley, most environmental groups support the assertion that a full restoration of the environment cannot be established without a radical overhaul of the capitalist system. In the Great Lakes area, no environmental group was willing to endorse this, with one exception: the umbrella organization Great Lakes United. This organization seems to be more 'radical', more purely egalitarian, than other environmental groups. As Great Lakes United is an umbrella organization and often speaks on behalf of other environmental groups, the egalitarian discourse seems to be somewhat over-represented in the Great Lakes regime.

Again, as in the Rhine case, it needs to be emphasized that there is quite a diversity of opinion among environmental groups. They may only be called egalitarian in comparison with firms and government agencies. Yet, some environmental organizations are more egalitarian than others. The discourse of the Environmental Defence Fund, for instance, has become mixed with individualistic elements and has started to advocate the application of market mechanisms as a solution to environmental problems.[31] The discourse of other environmental organizations contains a number of hierarchical notions.[32] For example, in March 1998 the members of the Sierra Club were invited to vote on whether to make it official Sierra Club policy to stop immigration into the United States.[33] This could hardly be called an egalitarian initiative. In fact, environmental organizations themselves are quite aware of their degree of 'radicalism' and can easily place themselves and other movements on an continuum depicting political radicalism. On this continuum Greenpeace and Great Lakes United are usually placed at the activist end of the spectrum, with Nature Conservancy, the Sierra Club and the National Audubon Society at the other.

The mindset and actions of the main governmental actor, the EPA, appear to mix egalitarian and hierarchical cultures. A number of egalitarian attitudes seem to drive the actions of EPA. The Great Lakes ecosystem is seen as fragile, and only 'zero discharge' of persistent chemical pollutants is seen as acceptable in the long run. Until then, the EPA favours very strict effluent limits and water quality standards. EPA also believes in a strict application of the precautionary principle. A great variety of chemicals is suspected to be toxic – many more substances are considered harmful by the EPA than industrial firms would agree to.

The egalitarianism of the EPA is not restricted to its view of the Great Lakes ecosystem. It also shines through in the agency's preferences with regard to public decision-making in the Great Lakes basin. These preferences cannot be estimated on the basis of the EPA's procedures for arriving at nationwide effluent limits and water quality standards which have been tightly prescribed by Congress. However, in the Great Lakes basin the EPA has had more leeway to structure decision-making processes, especially in its international negotiations with Environment Canada and in the development of the Remedial Action Plans. In both policy processes, the EPA has stimulated extensive involvement of environmental groups – an egalitarian ideal.

However, the EPA's dealings with enterprises have been decidedly hierarchical. In executing the Clean Water Act the EPA has followed a strict command-and-control approach. It has often taken decisions in great secrecy, without prior consultation with firms. It has usually prescribed in great detail which environmental goals corporations should meet, and has often laid down the precise technology to be used in meeting these standards. Sometimes, in addition, the EPA has strictly monitored and rigidly enforced its policies, meting out heavy fines on those firms which do not comply with the law. In sum, in its dealings with Great Lakes stakeholders, the EPA has combined an egalitarian preference for citizen participation and an egalitarian view of the environment with a hierarchical, top-down approach to its relations with the business sector. This forms a contrast with the ministries involved in the protection of the Rhine. The latter have tended to espouse more hierarchical views of the environment and the proper role of public participation, while having been somewhat less heavy-handed in their dealings with corporations.

There are two other major differences between the cultural configurations of the Rhine valley and the Great Lakes watershed. The evaluations of the environmental problems at hand have diverged less in the Rhine region than in the Great Lakes area. At the moment, the organizations in the Rhine basin agree that a major clean-up of the river has taken place. Environmental groups and governmental agencies feel that more could still be done, contrary to enterprises. So, various disagreements exist. Yet these differences of opinion pale in comparison to the controversies raging in the Great Lakes basin. There, environmental groups feel that the degradation of the ecosystem has not been halted, while firms are generally of the opinion that a major restoration has taken place. The views of government officials are usually somewhere in the middle of these two perceptions.

The same applies for the past. Over the last 25 years, Great Lakes firms have frequently disagreed with the EPA's evaluation of the extent of water pollution, as is evident from their testimonies before the US Senate as well as their attempts to have legislation repealed by decisions of the courts. At the same time, environmental groups have sometimes sued the EPA for not applying existing water protection laws strictly enough.[34] In the Rhine valley, there has been much less disagreement on the seriousness of pollution problems. From the late 1960s onwards, both industry and government have been concerned with the environment of the Rhine. The initial investments in sewage treatment plants by industries

coincided with a first flurry of governmental policies aimed at reducing pollution. Over the last 25 years, disagreements between
government and industry on the acuteness of the degradation of
the Rhine have of course been frequent. But these are small in
comparison with the gulf that has existed between the opinions of
agencies and firms on the seriousness of the pollution of the Great
Lakes.

Another important difference between the Rhine and Great Lakes
regimes concerns the extent of distrust and misunderstanding between organizations. In both areas, there has been a lot of mutual
hostility between business firms and environmental organizations.
In the Rhine valley this acrimony has become somewhat subdued
over the last five years, as the contributions of firms to the clean-
up of the Rhine have become better known. Much more severe
and lasting have been the miscomprehension and distrust between
industrial enterprises and governmental agencies in the Great Lakes
watershed. US firms complain that the EPA has not been concerned
with their profitability nor with the problems they have sometimes
had in meeting effluent standards. They also report that the EPA
has usually not been open to suggestions for cleaning up effluents
in other than the prescribed ways. Furthermore, enterprises are of
the opinion that the EPA's water policies have not been based on
'objective', convincing science. The EPA itself strongly disagrees on
this latter point. The agency asserts that all relevant scientific data
have been assessed in a fair way and suspects some firms of using
the impossible aim of absolute scientific certainty to stall water
protection. In general, the officials of the EPA feel that enterprises
will usually not make voluntary contributions to environmental protection, and that compliance of firms with existing laws needs to
be strictly monitored.

Both these differences between the cases (i.e. varying degrees of
distrust among actors, as well as varying degrees of convergence
on problem-definition among organizations) are central for understanding why Great Lakes firms have not been as cooperative on
pollution matters as their counterparts along the Rhine. Enterprises
will not want to make investments in water protection if they do
not think that a serious environmental problem exists. And they
will try to have laws watered down if they feel that these are not
based on sound reasoning, especially on 'sound science'. Great Lakes
corporations have opposed the EPA's water policies many times,
often challenging their scientific basis. These corporations have at-

tempted to have EPA policies repealed by the courts, Congress and the White House. During these appeals, it does not make sense for firms to start implementing the law pertaining to water protection. Any of their appeals might force a reversal in government policy, making certain investments in water protection unnecessary. In such a conflictual situation, it is out of the question that corporations would develop their own environmental policies (as firms have in the Rhine area). Firms that regard their legal obligations as too strict and unjustifiable will of course not attempt to go beyond the law. Furthermore, the constant reversals of government policy make it impossible to plan ahead for firms. If anything, companies will try to shirk their legal obligations to invest in water protection. This negative attitude by industry leads the EPA to develop laws that are even more rigid and comprehensive. The EPA will also start to mete out fines in order to bring the companies 'into line'. Both developments will create even more resentment among business executives. More rigid and comprehensive governmental policies will further close down the space for firms to develop their own environmental policies. In this adversarial process, environmental groups will lambaste firms for anti-social behaviour, and blame government for not clamping down on industry. In conclusion, distrust among actors, as well as widely divergent opinions about the seriousness of environmental issues, are central elements in an adversarial process that greatly diminishes the motivations of business leaders to allocate money to water protection.

Certain American institutions tend to lead to divisions between actors who look at environmental issues through different cultural lenses; or at least in comparison to a set of European institutions that tends to promote more of a rapprochement between alternative rationalities. An important part of the polarization process in America is the creation of distrust between organizations and the creation of widely divergent views of the seriousness of environmental issues. Below will be specified which American and European institutions have had these effects, and how these institutions have influenced the environmental politics of the Great Lakes and the Rhine. In particular it will be shown how differences in institutional contexts have influenced the motivations of corporations in the Great Lakes and Rhine basins to invest in water protection. This will provide a full answer to the puzzle of the present chapter.

A historical-institutional answer

It will be clear by now that institutions are conceptualized as those stable patterns of thought and practice that structure the relationships between the organizations involved in an issue-area that adhere to different worldviews (see also Chapter 3).[35] Grid-group theory is used to describe and clarify the content of these alternative worldviews. The theory could also be used to highlight the content and emergence of the institutions that regulate the links between the organizations and people involved in an issue-area. Grid-group analysis can thus be employed at different levels of analysis. However, when research is pursued on the developments within a single issue-area, it may not be worthwhile going to this trouble. A large part of the institutions impinging on the issue-area will typically have been formed in cultural clashes outside of the issue-area and in different periods than the one under review. Therefore, it may not always be feasible or efficient to employ grid-group theory for the purpose of finding the institutions that affect the actors within an issue-area. Instead, it may often be more economical either to rely on the existing comparative and institutional literature to come up with the relevant institutions and/or to do so in an inductive manner. The present chapter follows this second strategy, borrowing heavily from the literature on adversarial and consensual institutions.[36]

The historical version of the new institutionalism also treats institutions as the background variables that structure the relationships between the actors within a polity or economy in a given period. The analysis undertaken below can therefore be seen as falling under the purview of historical institutionalism, in particular, staying close to John Ikenberry's version of it. He distinguishes between several sets of relevant institutions, ranging from 'specific characteristics of government institutions, to the more overarching structures of state, to the nation's normative order'.[37] The explanation here of why the politics of water protection have been more adversarial in the Great Lakes basin than in the Rhine valley begins with the normative orders of the nations in which both watersheds are located. It is argued that there are enduring differences between the moralities that West Europeans and Americans tend to adhere to. These differences go some way in explaining the existence of adversarial water protection politics in the Great Lakes area. But not all the way, as the case of Switzerland shows. The normative order of the Swiss nation is exceptional among the Rhine countries, and is remarkably similar to the values adhered to by many Americans. Yet the Swiss have enjoyed highly consensual relations in the domain of water protection. This suggests that explanatory factors other than overarching moralities must have been important as well. These other factors are to be found at both the domestic and international level.

On the domestic plane, there are a number of differences between the state–society arrangements of the United States and those of the Rhine countries. These arrangements have shaped the relations between the executive, judiciary, parliament, business corporations and environmental groups quite differently on both sides of the Atlantic and have also tended to make US water protection politics more conflictual than the water protection politics within the Rhine countries. The actors in the Great Lakes and the Rhine watersheds have been severely affected by this. The Great Lakes basin has actually provided the major battlefield on which the struggles of American water politics have been decided. National business associations and environmental groups have frequently challenged the implementation of the Clean Water Act in the Great Lakes region. Moreover, Great Lakes companies and environmental groups have often taken the lead in shooting down US water legislation. References to these battles will follow. For now, it can be concluded that the state–society arrangements that have made US water politics adversarial have found their fullest expression in the politics of the protection of the Great Lakes.

These domestic institutional differences have been complemented by the differences among the international regimes for the protection of the Rhine and the Great Lakes. In particular the activities of the IJC have added fuel to the fires of the Great Lakes protection politics. International regime differences therefore form the last part of the explanation.

In sum, the historical-institutional account consists of three mutually supportive sets of arguments: two domestic (alternative moral orders and divergent state–society arrangements) and one international (regime differences). Together, they highlight the institutions that have made the Great Lakes protection politics more adversarial than the Rhine protection politics. This polarization in turn has made Great Lakes enterprises less willing to invest in water protection than Rhine companies. This explanation is clearly not parsimonious: it combines a variety of contingent, historically specific and interrelated explanatory factors. However, it fares far better than for instance the briefer explanation that would be offered by the rational choice version of the new institutionalism, as is shown later on.

The domestic level, part 1: American exceptionalism

More than other nations, Americans value liberty, egalitarianism, individualism, populism and laissez-faire.[38] Together, they form 'American exceptionalism' – a phrase coined by de Tocqueville in 1835.[39] Despite many other cultural changes, American exceptionalism has been in place for several centuries. One fiery passion underlies all elements of American exceptionalism: a dislike and distrust of central government. How can this anti-authoritarian thrust be linked to the adversarial nature of water protection politics in the United States, including the Great Lakes states? The answer is: in at least two ways.[40] The story begins with the growing concern for the environment in the late 1960s. This development did not of course only take place in America, but elsewhere as well. What was different in America was that the demands for a cleaner environment were not accompanied by a desire to expand government. The anti-statism within both environmental and other organizations prevented this. Thus the paradoxical situation occurred in which the US government was implored to restore the environment without receiving sufficient means to do so. This paradox has hobbled the implementation of the Clean Water and Air Acts. During the formulation of both acts, environmental organizations effectively lobbied Congress and as a result both Acts included ambitious environmental standards. The Acts also increased the budget of the EPA, but not enough to realize the unrealistically strict environmental goals. This process set off a vicious circle, one of several that have kept American environmental politics adversarial. Forced to achieve impossibly strict aims with not enough means at its disposal, the EPA could only fail. This process kept alive the belief that bureaucracy is not capable of getting things done and should remain limited. Businesses, also faced with impossibly high demands on them, tried to get legislation repealed. This added to the suspicions of industry harboured by environmentalists and increased their calls for stricter controls. It also induced the EPA to be 'tougher' on industry, which further increased the resentment against government among business executives. Setting off these polarizing processes is one way in which

American exceptionalism has contributed to adversarial environmental politics in the United States.

Another way is by holding up an ideal of rugged individualism. 'Standing up for yourself' and 'holding your own' are cultural traits that are especially valued in the United States, but such beliefs are not particularly conducive to dialogue among organizations with divergent views of environmental issues. They also do not stimulate acceptance of government policies. As Huntington writes:

> The ideological pluralism in Europe also means that liberal, democratic, and egalitarian norms are generally weaker in European countries than they are in the United States and that nonliberal, nondemocratic norms stressing hierarchy, authority and deference are stronger. Comparisons of political culture consistently document these differences.[41]

American exceptionalism has therefore been an important cause of the antagonisms that have consumed environmental politics in the United States. This is all the more so since the anti-authoritarian values of American exceptionalism have most fervently been adhered to within the US business community. David Vogel has shown this, saying 'the most characteristic, distinctive and persistent belief of American corporate executives is an underlying suspicion and mistrust of government', and that actually 'businessmen are more anti-statist than virtually any other major interest in American society'.[42] This anti-statism has also prevailed among Great Lakes business people. In their eyes, water protection should only proceed on a voluntary basis, and should not be imposed by government.[43] As American exceptionalism has also been believed in by Great Lakes business leaders, it can be said that this set of norms and values has been one source feeding the antagonisms between the actors in the Great Lakes basin.

Is American exceptionalism enough to explain the adversarial nature of water protection politics in the Great Lakes area? The case of Switzerland shows that this is not so. Swiss society is characterized by a normative order remarkably akin to that of the United States: the Swiss have opposed central authority as intensely as have US citizens. This is apparent in many ways.[44] To name just one example, the Swiss government has never had much involvement in the economy.[45] Yet, despite a comparable normative order, Swiss environmental politics have been highly consensual. This shows that it is possible to combine an anti-hierarchical morality with consensual environmental politics. An anti-authoritarian stance is therefore not sufficient for the emergence of adversarial environmental politics. Such a moral order is only one element that goes into the making of adversarial water protection politics. Other elements are also needed which often consist of the historical institutions that have shaped the relations between the executive, judiciary, legislature, business organizations and environmental groups in the involved countries. These domestic institutions have been organized quite differently in the United States than in the Rhine countries. As a result, water politics have been more antagonistic in the United States than in Western Europe. Below is described how this has been possible, as well as how this has affected the protection efforts regarding the Rhine and the Great Lakes.

The domestic level, part 2: state–society arrangements

The executive and judiciary

The antagonisms among organizations involved in American water politics have been fuelled by the ample opportunities that non-governmental actors have had to challenge laws and administrative policies through the courts.[46] In Europe these opportunities have been smaller. What follows first describes when US citizens and organizations can request judicial review, and shows that these opportunities have been more limited in the Rhine countries.[47] Then comes a consideration of how these differences have made American water politics more conflictual as compared to the European situation, and finally how all of this has affected the Great Lakes and Rhine basins is described.

One way in which Americans have reined in governmental power is by offering courts of law the possibility of constitutional review. In the United States the judiciary has the right to annul laws that have been adopted by Congress on the grounds that they are unconstitutional, and US private citizens and organizations can ask courts to rule on the constitutionality of legislation. In the Rhine countries, the opportunities for private actors to request constitutional review are more limited, as the primacy of parliament in deciding on the adoption of laws is seen as a fundamental democratic principle. Administrative review is another element that gives US judges a bigger political role than their European colleagues. American courts have been allowed to rule on how governmental agencies have implemented Acts (and especially environmental Acts) adopted by Congress. Both citizens' organizations and companies have had standing in these courts. Again, the situation is different in the Rhine countries where the role of legal courts in administrative review is more limited.

Even more important than the formal provisions for judicial review are the actual practices, habits and values of legal actors.[48] In the United States, an active political role by the courts has been widely viewed as legitimate and necessary. Judicial review is an important element in the American system of checks and balances that aims at limiting central authority.[49] By contrast, in Europe the power to make or repeal legislation has usually been seen as a parliamentary prerogative.

All of these legal practices and attitudes make the political role of the judiciary in the Rhine countries smaller than in the United States. How has this contributed to more adversarial environmental politics in America? First, American corporations simply do not have to accept the EPA's protection policies. They can always seek reversal of these policies in court, and for the duration of these court cases, firms can withhold investments in environmental protection. This is much resented by the EPA, which also runs the risk of being sued by citizens' organizations for not meeting environmental goals on time. The EPA's water protection policies have frequently been challenged in the courts, and while these legal challenges have often not led to reversals of the EPA's water protection policies much valuable time has been lost during court proceedings.

Furthermore, court cases themselves are not exactly conducive to a coming together of minds. During law suits it is beneficial for parties to present their views as strongly as possible and it is counterproductive to show any sympathy for the opinions of the opposite party. In court, business representatives will testify that there is no scientific base whatsoever for the EPA's protection policies. EPA officials will argue the exact opposite.

For these reasons, the institutions that have regulated the relations between the judiciary and the executive in the United States have contributed to the disagreements among public and private organizations that have characterized American environmental politics, including the politics of water protection.

The Great Lakes basin has been at the centre of this legal haggling. To begin with nationwide business associations have often sought to influence US water policies through court cases dealing with the implications of national laws for the Great Lakes basin.[50] It should not be surprising that national business associations have frequently focused on the basin. As over half of the 500 largest US industrial companies are located in the watershed, Great Lakes firms have had a major say in such business associations as the American Automobile Manufacturers Association, the American Iron and Steel Institute, the American Forest and Paper Association and the Chemical Manufacturers Association. Furthermore, Great Lakes companies themselves have often legally challenged US water laws.[51] Great Lakes enterprises have therefore both directly and indirectly led much of the legal opposition of business against the Clean Water Act. Much the same can be said of American environmental organizations. They too have frequently looked upon the Great Lakes as a major battleground on which to decide the fate of US water legislation, and both local and national environmental organizations have often sued the EPA for not having enforced the Clean Water Act strictly enough in the Great Lakes basin.[52] Environmental groups (especially the Atlantic States Legal Foundation) have also indicted Great Lakes firms for allegedly not having complied with wastewater legislation.[53] These court cases illustrate that the Great Lakes have been at the very centre of the water protection battles between American firms, government departments and citizens' groups. The ample opportunities that these actors have had for suing each other have deepened the rifts between them.

The executive and legislature

The institutions shaping the relations between the executive and the legislature in America have also been different from those in the Rhine states. In particular, in the United States there is a presidential system, while most of the Rhine countries have a parliamentary system.[54] More accurately, the United States has a 'pure' presidential system, the Netherlands and Germany have 'pure' parliamentary systems, while the government systems of France and Switzerland are hybrids of these models. A presidential system displays the following characteristics: (a) the executive and the legislature are separately chosen by the public; (b) the government cannot be forced to resign with a parliamentary vote of no confidence; and (c) executive power is concentrated in one person. A presidential system is largely based on the ideal of a separation of powers: it pits government against parliament. A parliamentary system has the following traits:

(a) the government is chosen by the popularly elected parliament; (b) the legislature can force the government to resign; (c) executive power is exercised in a collegial manner, i.e. ministers take decisions jointly. In a parliamentary system, the executive and legislature cooperate more closely.

In several ways presidential systems stoke the fires of environmental politics higher. First, in a presidential system parliamentarians bear no responsibility for, and therefore tend to be less concerned about, the implementation of legislation. In fact, if government fails to achieve the aims of legislation, the latter acquires a stick with which to beat the former. Thus it can be an attractive strategy to adopt laws that incorporate impractical, extreme measures, particularly when the parliamentary majority belongs to a different political party from the executive. Such lack of parliamentary responsibility was taken to extremes with the adoption of the US Clean Water Act by the Democratic majority in Congress in 1972. This Act included 'stringent timetables that ranged between the merely unrealistic and the wholly fantastic' – against the will of President Nixon.[55] As a consequence, Great Lakes firms were set against the Clean Water Act right from the start.[56]

Second, a presidential system gives interest groups another opportunity to challenge government policies. In presidential systems there tends to be a struggle for power between the executive and legislature. In the United States, Congressional committees try to keep a tight rein on government departments by holding frequent reviews of their policies. Private organizations, such as business associations and citizens' groups, lobby members of Congress and testify before the various parliamentary committees. Thus these groups have another chance to get government decisions repealed, which tends to increase the acrimony between governmental and non-governmental actors.[57] These processes have affected American water protection politics as well. The Clean Water Act has been regularly reviewed by the US Senate. During the reviews, environmental groups, business representatives as well as EPA officials have offered their opinions. As with court cases, Senate hearings induce organizations to present their views as strongly as possible – thus deepening their divisions. Companies and environmental groups from the Great Lakes watershed have played a dominant role in Senate hearings on water legislation. Both camps have frequently offered their opposite viewpoints in strongly worded testimonies to the Senate.[58] Under parliamentary systems, private actors have fewer opportunities to influence government policy through the legislature, as in these systems public power is concentrated more in the executive than the legislature.

A last argument relates to executive–executive relations. In parliamentary systems, policy decisions are typically taken after a lengthy process of consultation among various ministries. This interministerial consultation reassures private actors that their wishes will be considered. In both Germany and Holland, the Ministry of Economic Affairs has defended the standpoints of business associations in the formation of domestic and international policies concerning the Rhine. Likewise, the Dutch and German Environment Ministries have usually sided more with environmental groups in the formation of Rhine policies. This has further reduced the need for private actors within Germany and the Netherlands to agitate against government policies affecting the Rhine. In presidential systems, government agencies develop their policies in relative isolation from each other, and then ask for permission to implement these policies from the head of government. The EPA seldom negotiates with other government agencies over its proposed effluent limits, and concentrates instead on how to sell its strict guidelines to the White House. This has also polarized American and Great Lakes protection politics.

The above arguments are based on the fact that the polity of the United States has resembled a pure presidential model, whereas those of the Netherlands and Germany have approached a pure parliamentary model. These arguments apply less to Switzerland and France, as the polities of these countries are hybrids of the two types. The Swiss polity combines two characteristics of the parliamentary model with one property of the presidential model.[59] The first two are collegial government and parliamentary selection of the government. The latter is the fact that the Swiss government is not dependent on legislative confidence. But this one deviation from the pure parliamentary model only strengthens the Swiss executive *vis-à-vis* the legislature. It certainly does not pit parliament against government, nor does it offer private actors any chance to seek redress of government policies in parliament.

The French political system of the Fifth Republic is Janus-faced. When the political party of the president holds a majority in the Assemblée, the French system resembles the parliamentary model. When the parliamentary majority opposes the president, the French polity comes closer to the presidential model.[60] But even during its presidential phases, French environmental politics are much less affected by controversies between government and parliament than in America. This is because the constitution of the Fifth Republic incorporates various measures that sharply curb parliamentary control over the government.[61] As a result, French interest groups cannot reasonably hope to change governmental policies through pressure on the Assemblée.

The executive and business corporations

A missing link in the explanation concerns the institutions that shape the direct contacts between government departments and corporations in the United States and the Rhine countries. These links have been organized quite differently on both sides of the Atlantic. Below are first considered the effects on both watersheds of the general ways in which business–government relations have been organized in the two areas. Thereafter, the impact on the two basins of business–government relations in the field of environmental protection is considered.

In many European countries, corporatism has long reigned. According to Philippe Schmitter, corporatism is

> a system of interest representation in which the constituent units are organised into a limited number of singular, compulsory, non-competitive, hierarchically ordered and functionally differentiated categories, recognised or licensed (if not created) by the state and granted a deliberate representational monopoly within their respective categories in exchange for observing certain controls on their selection of leaders and articulation of demands and supports.[62]

In a corporatist economy, private actors are politically represented by a limited number of relatively stable and large interest associations. Firms are represented by employers' organizations, employees by labour unions. Their negotiations take place under the watchful eye of the government. The interest associations typically try to find a consensus among themselves that is acceptable to their own members and the government. As such, interest associations in a corporatist economy often 'deputize' for the executive. The corporatist model is essentially hierarchical and corporatists base 'their faith either on the superior wisdom of an authoritarian leader or the enlightened foresight of technocratic planners'.[63]

The specifics of the corporatist model have differed from country to country and from time to time.[64] Despite these fluctuations, the economies of Germany, the Netherlands and Switzerland have usually been seen as clear examples of the corporatist model, certainly since the end of the Second World War.[65] France, on the other hand, does not have a corporatist system of interest representation.[66] The French state seems too dominant and interventionist for a corporatist system to work well, while the unions and business associations have often been unable to reach consensus.

How does this tie in with the voluntary investments in water protection made by firms in the Rhine area? The argument is certainly not that environmental politics in the Rhine basin have proceeded along corporatist lines (with environmental organizations taking the place of labour unions). This has certainly not been the case (see below). The effects of corporatism on the protection of the Rhine have been indirect. In all Rhine countries (except France) corporatism has familiarized companies with a form of self-regulation. Under corporatism, firms voluntarily agree to take measures that are viewed as desirable or acceptable by the executive. This is exactly what has taken place along the Rhine river. Firms have made investments in water protection that were not required by the government but were still deemed desirable by the latter. In addition, corporatism has created the organizational means through which industry-wide environmental policies can be coordinated. The business associations that are an essential part of any corporatist system can fulfil such a role. This has been especially important in Germany. The German Association of the Chemical Industry (VCI) has been a motor behind the massive investments in water and air protection undertaken by the chemical firms in Germany.[67] For instance, the VCI has been instrumental in the development of informal industry-wide effluent norms. Moreover, it has set up a programme through which the large chemical concerns have helped smaller firms to purify their discharges. The relevance of the VCI's activities for the clean-up of the Rhine becomes clear when one considers that the largest German chemical firms are located on the waterway.

Corporatism has never taken root in America, being too much based on hierarchical principles.[68] American business leaders have been too independent to accept the formation of authoritative interest associations. Similarly, only a small percentage of American workers have been willing to be represented by a union – a much smaller number than in the Rhine countries. Instead, the political economy of the United States has been characterized by pluralism. Schmitter defines pluralism as

> a system of interest representation in which the constituent units are organised into an unspecified number of multiple, voluntary, competitive, non-hierarchically ordered and self-determined (as to type or scope of interest) categories which are not specially licensed, recognised, subsidised, created or otherwise controlled in leadership selection or interest articulation by the state and which do not exercise a monopoly of representational activity with their respective categories.[69]

Under pluralism, individual actors (be they firms or persons) fend for themselves. They form temporary alliances with other actors if this seems to further their self-interest. But the alliance immediately unravels, the moment it no longer serves this self-interest. Thinking and acting in terms of a 'group interest' is much less developed under pluralism than in a corporatist system. Under pluralism, decision-making is highly fragmented. The system rests on the belief that those actors for whom the outcome of a certain issue matters most will usually spend more resources on influencing decision-making regarding the issue than other actors. As a consequence, so the assumption continues, issues will most often be decided in favour of those actors for whom they matter most (at least when power resources remain evenly distributed).[70]

There are two ways in which the US pluralist system of interest representation helps to explain why American firms in the Great Lakes watershed have balked at making investments in water protection. First, a pluralist system induces actors to think in individualistic, rather than social, terms. Investment in water protection undertaken by a firm benefits all those who live in the watershed. The costs of the investment, however, fall disproportionately on the firm. From a narrow self-interested point of view, this is a problem. Standard economic analysis (based on the assumption of self-interested behaviour) would predict under-investment. However, if actors thought more in social terms there would be less under-investment. Therefore, to the extent that a pluralist system strengthens thinking in terms of narrow self-interest, it also diminishes firms' willingness to invest in environmental protection. Second, under pluralism the organizational basis for voluntary industry-wide water protection programmes is lacking. In the Great Lakes basin, no central industry association has had enough authority to induce firms to invest in water protection programmes. The Council of Great Lakes Industries only started work in the early 1990s, and has remained a tiny organization, capable of representing the Great Lakes firms in political fora but not strong enough to influence the environmental stances of the companies.

The manner in which water protection policies have evolved in the Rhine countries is a long haul from the extensive concordance between public and private organizations that is characteristic of corporatism.[71] By and large the Rhine states have used 'command-and-control' approaches to environmental protection.[72] They have attempted to force industries to invest in aquatic protection by adopting water quality standards and effluent limits. In principle, Rhine firms should have been able to obtain discharge permits only if their effluents met the legally required standards. The politics of water protection within the Rhine countries have only been 'consensual' in comparison to the extremely adversarial water protection politics in the United States.

In the United States, the EPA has also opted for a strict regulatory policy approach.[73] But not every command-and-control policy equals the other. Clearly, the EPA's approach to water protection has been much more rigid, top-down and legalistic than the water policies adopted in the Rhine countries.[74] For example, industry's views have carried less weight in the EPA's decision-making processes than has been the case in the Rhine valley. Moreover, American emission values and technological standards have been more detailed and stricter than European ones. Lastly, in the Rhine countries firms temporarily unable to fulfil their legal obligations towards the environment have sometimes been able to discuss this problem with the authorities.[75] Government officials from these countries have certainly not threatened firms with severe penalties. In similar cases in the United States, the EPA has shown little understanding. Instead, it has frequently handed out heavy fines to firms and sometimes even sought imprisonment for business executives who have not met environmental standards.[76] Again, the Great Lakes ecosystem has been in the midst of all this action. The EPA has often sued Great Lakes firms for not having complied with national water legislation.[77]

Another factor has also made the command-and-control policies of the Rhine states less oppressive than the command-and-control policies of the EPA. In two Rhine countries the governments have ceased favouring a command-and-control approach to water protection. From the mid-1980s onwards, the Swiss and Dutch governments have put more and more emphasis on voluntary programmes.[78]

A last development that has softened the impact of the command-and-control policies in the Rhine countries concerns their actual implementation. In both the Netherlands and France a gap has existed between the way in which water protection has been formulated and the way in which it has been implemented. The policy-*makers* in both countries have preferred strict regulation of polluters. They have formulated national norms to be implemented in stringent protection policies. But the policy-*implementers* in both France and the Netherlands have often diverged from these strict controls. In the Netherlands, the civil servants responsible for implementing water policies have realized that they simply do not have the financial means to systematically control the emissions of firms. Moreover, they have experienced great difficulties in providing evidence of environmental infringements. For these reasons, Dutch policy-implementers have opted for a more flexible approach than the strict command-and-control policies favoured by central policy-makers.[79]

In France, the local *préfet* has been responsible for the issuing of discharge permits and the implementation of centrally developed environmental policies. The *préfet* is not only the local representative of the Ministry of the Environment, but also of other governmental departments. As such, he or she is expected to strike a balance between the opposing interests of the various departments. For instance, on the one hand the *préfet* has to implement water protection policies in his or her region, but on the other hand he or she also has to take into account the local needs for industry and employment. Legally, the *préfet* can diverge from national environmental norms on the grounds of these other public needs. Moreover, the *préfet* is not held to enforce national water quality standards if in his or her view the local environmental conditions do not require this. As a result, in France the centrally adopted environmental standards have often become watered down locally.[80] This has played an important role in the *basin de l'eau Rhin–Meuse* (the administrative region of France through which the Rhine flows). The *préfet* responsible there has been quite sensitive to the needs of local industries. For instance, when on various occasions in the 1980s

French courts ruled that an effluent permit given to a major Alsatian mine company should be withdrawn, the *préfet* was quick to do so, and even quicker to issue a new, equally broad permit to the company.[81]

Command-and-control policies are usually resented by firms. The strict regulatory approach with which EPA has tried to force Great Lakes firms to comply with the Clean Water Act has therefore contributed significantly to the distrust that has characterized the environmental protection of the Great Lakes.

The executive and environmental organizations

The room for influence that environmental groups have had in the Rhine valley has differed from country to country. In Switzerland, environmental groups have had the greatest opportunities for effecting policy change, in France the least.[82] However, any opportunities for influencing government policies that environmental groups have had in Western Europe have been dwarfed by those in America. Besides the challenges that environmental groups can launch through the courts, Congress and the White House, the EPA has offered environmental groups ample opportunities to directly affect its policies. Formally, the EPA is obliged to invite citizens' groups to comment on proposed legislation. Moreover, informally, the EPA has often favoured the viewpoints of environmental organizations over those of corporations. Unsurprisingly, the business community has felt 'exposed' by the influence environmental organizations have sometimes been able to exert on the EPA and has sought to fight back.[83] In the Rhine countries, government officials have been less impressed with the arguments of environmental groups and have tended to be more 'neutral' – thus easing the qualms of corporations. The institutions that have regulated the contacts between the EPA and American environmental organizations have therefore been an additional source of the distrust and disagreement that have plagued the protection politics of the Great Lakes – next to the institutions that regulate the relations between the executive, judiciary, legislature and business community.

Companies and companies

Thus far the argument have been made that alternative state–society arrangements have made the politics of water protection more adversarial in the Great Lakes basin than in the Rhine watershed which has reduced the willingness of Great Lakes firms to invest in water protection. These arguments should be complemented by a consideration of the regional variation of the relations within and between companies. This variation is the topic of the 'capitalism versus capitalism' literature, which usually distinguishes between two rival forms of capitalism.[84] The one kind stresses a consensual, group-oriented organization of the economy (sometimes called 'the Rhine model') and is most often seen as exemplified by Germany and Japan, and to a somewhat lesser extent by Holland and Switzerland as well. The other capitalism ('Anglo-Saxon capitalism') is based more on individualistic and antagonistic principles and is represented most by the United States.

This literature highlights two differences among the business communities of the United States and the Rhine countries that are relevant for the puzzle of this chapter. The first difference concerns the relations between industrial concerns and their financiers. US firms raise money more on the stock markets, whereas firms in the Rhine countries rely more on self-financing and bank loans. This makes US companies more beholden to stockholders, who often have less of a stake in the survival or social standing of the company and are usually more interested in the current value of their earnings. To keep their stockholders happy, US firms are therefore induced to maximize short-term financial profits. Particularly since the 1980s, this tendency has been strengthened by the phenomena of 'corporate raiding' and 'hostile takeovers', involving the involuntary takeover of a firm (through the stock market), usually followed by the sacking of its management. These phenomena have focused the minds of US corporate leaders even more on short-term profits. Water protection, which involves long-term planning and financial sacrifices, begets a lesser priority under these circumstances.

In the Rhine countries, firms finance their activities more through large banks than via the stock market. Moreover, the shares that they *have* issued have often ended up in the hands of the very same banks. The large banks in the Rhine countries (and especially in Germany) have therefore occupied a central role in the functioning of enterprises. These banks have tended to believe that the profitability of companies depends on long-term planning and often entails short-term losses. They have been much less willing to sell and buy domestic assets at short notice. Thus, the banks have sheltered firms from hostile takeovers. Furthermore, particularly in Germany, banks and companies from various sectors have grouped together – holding onto each other's shares and providing financial help in hard times. Many German companies from the Rhine basin have been part of such corporate groups,[85] whereas in while in the Netherlands and Switzerland such groups are less prevalent. Dutch firms have been protected from hostile takeovers by extensive legal prohibitions and Swiss firms by restrictive rules governing share-holders' voting rights. All of this has freed Rhine companies somewhat from a single-minded pursuit of short-term profit. This has also eased the making of extensive voluntary investments in water pro-tection in the Rhine basin, as these investments have entailed long-term planning and reduced profits. In particular, the water protection efforts of the chemical giants Bayer AG and Hoechst AG seem to have been facilitated by this lesser dependence on the stock market, as both these concerns have taken great risks in their efforts to revolutionize wastewater purification technology. Bayer's 'tower biology' and the 'Bio-Hochreaktor' of Hoechst made reductions in water pollution possible that had hitherto been unthinkable. It seems unlikely that these enterprises could have taken the huge risks in-volved in developing their technologies if they had not been sheltered from hostile takeovers and shareholders mainly interested in short-term profits.

A second relevant difference among the rival capitalisms concerns the relations between management and employees. Again, in the Rhine countries (except France) these relations have been more convivial and oriented towards the long run than in the United States. In America, labour tends to be seen as just another com-modity. It is offered and sold without much ado. This leads to a high degree of job rotation in the United States, which in turn tends to make work relations within US companies more imper-sonal and geared toward the short term.[86] One manifestation of this is the lower rate of on-the-job training in American firms than in (especially) German companies. Another manifestation is the far greater influence that employees have on company policy in the Rhine countries. In the Netherlands, Germany and Switzerland, the concerns and views of employees loom larger in the decision-making of firms than in America. This latter process has been important for the clean-up of the Rhine. Asked why their firms had invested in water protection, the representatives of Dutch, German and Swiss companies in the Rhine watershed offered another motivation besides stating their wish to do the (socially) right thing. They said that their employees (most of whom live in the watershed) were also in favour of such a course. So, the greater willingness of corporate leaders to take into account the views of their employees in the Rhine model of capitalism constitutes an important part of the solution to the riddle of this chapter.

In sum, the less rivalrous relations *between* Rhine corporations (facilitated by a different financing system), as well as the more convivial relations *within* Rhine companies, have been conducive to long-term investments in the protection of the river. As this 'rival capitalisms' argument is ultimately also based on the difference between adversarial and cooperative social relations, this proposition is considered to be complementary to (rather than competing with) the arguments derived above from other state–society differences.

The international level: the role of the International Joint Commission

The third and last part of the explanation of the puzzle concerns the international level. The international regime for the protection of the Great Lakes has diverged in several respects from the international regime for the Rhine. These regime differences have especially been shaped by the International Joint Commission, and have amplified the antagonisms among the organizations involved in the pollution and protection of the Great Lakes. As such, the international regime for the Lakes has further widened the rifts among actors that have been created by domestic institutions.

In a number of ways the IJC has stoked the fires of the environmental politics of the Great Lakes even higher, firstly through its elaborate public participation programmes. Every two years, the IJC brings out a new 'Biennial Report on Great Lakes Water Quality'. In these reports the IJC comments on the ongoing implementation of the 1978 Great Lakes Water Quality Agreement by the US and Canadian governments. Before the IJC writes a Biennial Report, it organizes a Biennial Meeting at which stakeholders can present their views on the protection of the Great Lakes to the Commission. Usually, thousands of representatives from non-governmental organizations attend these meetings. The opinions of persons attending these conferences often find their way into the Biennial Reports. Sometimes, the IJC also invites non-governmental actors to serve on its advisory boards. The IJC is very proud of its public participation processes. They are indeed unique in the world. However, this public participation has seldom consisted of input by corporate leaders.

Representatives from firms only started to come to the Biennial Meetings from the beginning of the 1990s. At both the 1991 and 1993 conferences, they were shouted down by a hostile crowd of environmentalists – an experience they did not particularly relish. Business representatives have also seldom served on the advisory boards of the IJC.

A second way in which the IJC has polarized the Great Lakes regime is through its recommendations to the North American governments. The 1978 Great Lakes Water Quality Agreement includes the goal to 'virtually eliminate' industrial emissions of persistent toxic substances. A whole political battle has been waged over the interpretation of the concept of 'virtual elimination'. Environmental groups have advocated a strict interpretation of this concept, asserting that 'zero means zero'. Great Lakes firms have rallied against such a narrow interpretation. They have contended that such a strict interpretation would not only threaten their businesses, but would also not be a cost-efficient form of protecting the Great Lakes. In their eyes, greater ecological gains are to be had by investments elsewhere. The IJC has aligned itself with the environmental movements. Through its publications it has aggressively exhorted the governments to strive for a full elimination of persistent toxic substances.

Another important issue has been the fate of chlorine in the Great Lakes basin. In a hotly contested recommendation to the North American governments in 1992, the IJC called for the elimination of chlorine from the Great Lakes ecosystems.[87] This recommendation has infuriated Great Lakes firms and served as a lightning rod for their opposition against the IJC. Both in the United States and in Canada, corporations have established organizations with the sole task of blocking the acceptance of this one IJC recommendation by the national governments: the Chlorine Chemistry Council in Washington, DC, and the Canadian Chlorine Coordinating Committee in Ottawa. These organizations have had a rather powerful argument: chlorine is an element from the periodic table. An unsympathetic reading of the recommendation by the IJC is therefore that the international organization has proposed to remove a natural element from the Great Lakes ecosystems. This one recommendation, more than anything else, has badly damaged the IJC's reputation among business executives. It has given them powerful ammunition to claim that the IJC is partisan and bases its proposals on 'bad science'. In fact, the chlorine recommendation of 1992 sparked a debate among Great Lakes firms as to whether to lobby for the abolition of this 90-year-old international organization altogether.

In sum, the International Joint Commission has taken a rather one-sided view of the environmental issues in the Great Lakes area. This is fully understandable within an adversarial setting, and (within that setting) probably beneficial for the environment. However, the IJC's actions have at the same time deepened the rifts that already existed between the actors in the Great Lakes regime. Thus, the IJC has reinforced the adversarial system of water protection in (the United States side of) the Great Lakes area. This need not have been inevitable, as the IJC has also had first-hand experience with a more consensual method of water protection, namely that of Canada. So, the IJC has missed a chance to bring firms, environmental groups and government departments closer together. This would also seem to have been harmful to the environmental protection of the Great Lakes, as ultimately the adversarial relations that have existed between Great Lakes firms, environmental organizations and the EPA have been the main cause of the (relative) lack of willingness among Great Lakes corporations to invest in water protection.

Which institutional answer is the right one?

The puzzle of this chapter has been answered by combining the historical version of the new institutionalism with grid-group theory. But this leaves out another possible institutional answer: the one that might be provided by the rational choice version of the new institutionalism.[88] Here, the extent to which this alternative institutional framework could solve the puzzle of this chapter is considered.

A central concern of the rational choice institutionalism is the principal–agent problem. This states that a main worry of 'principals' consists of ensuring that the 'agents' whom they have employed will implement the policies that the principals have decided upon instead of acting upon their own preferences and interests. Principals can never be sure that agents do not deviate from the plans they have set, since (in a world of costly information) they have much less information about the behaviour of their agents than do the agents themselves. The rational choice literature has highlighted two mechanisms that may overcome this problem: (a) administrative procedures; and (b) oversight procedures.[89] The former limits the agents' autonomy by precisely formulating their tasks, working methods and resources. The latter achieves the same by monitoring and rewarding as well as sanctioning agents. Indeed, McCubbins and Schwartz have distinguished between two forms of oversight: police controls and fire alarms,[90] the difference being that police controls consist of monitoring by the principals themselves whereas fire alarms are set off by third parties (usually interest groups). Most of this literature has been applied in studies of how parliaments attempt to limit the autonomy of regulatory agencies.[91] However, the implications of this model for the relationship between government agencies (as principals) and the organizations that they attempt to regulate (the agents) are clear as well: government agencies should attempt to constrain the autonomy of those that they want to regulate by engaging in a mix of administrative controls and oversight mechanisms.

Could this rational choice literature solve any part of the puzzle in this chapter? The answer is that it certainly could not. In fact, the insights provided by rational choice only add to the mystery of why the industrial effluents into the Rhine have been less toxic than the discharges into the Great Lakes, for both administrative and oversight procedures have been used to a far greater extent by the environmental authorities in the Great Lakes basin than in the Rhine watershed. *Regarding administrative procedures*, the US water protection laws pertaining to the Great Lakes area (such as the 1972 Clean Water Act and the 1990 Great Lakes Critical Program Act) have been much more specific and extensive than the water protection laws prevalent in the Rhine countries. The US regulations have laid out overall policy goals, water quality standards, effluent norms and permitted technologies in a more comprehensive way than the European water laws. For example, while the current German effluent standards take up only a few pages, the American effluent standards at present take up 1592 pages of the US Code of Federal Regulations. It can therefore be concluded that the American water protection laws have prescribed the activities of firms to a greater degree than European laws have. *Concerning police controls*, as has already been discussed, the US authorities responsible for the protection of the Great Lakes have been more heavy-handed than their counterparts in the Rhine countries. The EPA has tried to watch closely how state authorities and firms have implemented its water protection policies, and has not hesitated to sanction states and enterprises that have fallen behind. Such direct monitoring and severe sanctioning have been much less prevalent in the Rhine countries, where implementation of water protection laws by regulatory agencies has often been half-hearted. Finally, *with respect to fire alarms*, these have been extensively used in the environmental protection of the Great Lakes. The American and Canadian governments have used the International Joint Commission as a fire alarm with respect to the clean-up of the Lakes. Every two years, the IJC publishes a thick and critical report on the progress that the two governments have made under the 1978 Great Lakes Water Quality Agreement. In these biennial reports, the views of many environmental groups on the clean-up of the Great Lakes are included. These international fire alarms are complemented by domestic ones. Under the 1986 Citizens-Right-to-Know Act, US companies are obliged to report the content of their releases into the open air and water. Moreover, environmental groups have the right under US law to sue the EPA and firms for not implementing water protection policies swiftly enough. Such fire alarms are much less in use in the Rhine countries. It must therefore be concluded that the US agencies responsible for the protection of the Great Lakes have wielded the weapons of administrative procedures and police controls as well as fire alarms to a far greater extent than their counterparts in the Rhine countries. But even with all these efforts, the industrial discharges into the Rhine have been less toxic.

Where does the rational choice variant of the new institutionalism go wrong? This literature paints a world populated by calculating and purely self-interested actors. In this world, institutions merely distribute power resources among clashing actors. It omits the effect that institutions may have on the mindsets of actors. The comparison here of the environmental protection of the Rhine and the Great Lakes suggests, however, that institutions that are predicated on the existence of purely self-interested actors may actually help to sustain exclusively self-centred attitudes among actors. That is what rational choice fails to consider.

What might work best?

To explain why the industrial effluents into the Rhine have been cleaner than the releases into the Great Lakes, three steps have been taken. First, it was shown that Rhine firms have made extensive voluntary investments in water protection, whereas Great Lakes enterprises have restricted their protection efforts to a bare minimum. Then it was argued that these divergent inclinations to invest in water protection have sprung from two alternative ways of conducting environmental politics: the more adversarial processes in the Great Lakes region, and the relatively (!) consensual processes in the Rhine area. Lastly it was explained where these alternative methods of conducting environmental politics have sprung from: American exceptionalism, different state–society arrangements in the two regions and international regime differences. One issue still remains: in general, can it be said that more cooperative institutions lead to more comprehensive environmental protection than institutions that polarize? The issue seems clear-cut, judging by the evidence presented here. The more consensual politics of the Rhine countries have led to cleaner industrial effluents than the antagonistic policy processes in the Great Lakes watershed. Yet a number of reservations must be made. First, it needs to be remembered that only two environmental regimes have been compared, which does not allow for overly strong inferences. In addition, the focus was on only one circumscribed aspect of these two regimes, namely industrial discharges into water. Industrial discharges by air, polluted sediments, agricultural emissions or the loss of habitat have not been considered, and it is not clear how including any of these factors would have affected the conclusions.

It also important to realize that the research has reached different conclusions than four other comparative studies of environmental politics on both sides of the Atlantic: those of Vogel (1986), Badaracco (1985), Wilson (1985) and Brickman, Jasanoff and Ilgen (1985).[92] These studies compare various other aspects of environmental politics in the United States with regulation in Western European countries. They reach quite similar conclusions concerning the 'independent variable' of this study, namely that America's political institutions give rise to much more adversarial environmental policy processes than European institutions. But they reach dissimilar conclusions regarding the 'dependent variable', i.e. the level of environmental protection that has been achieved. They estimate that in their cases the environmental protection that was achieved in the United States roughly equalled that of European countries. So, in their cases, similar environmental results were reached via two contrasting ways: an adversarial path in America and a more consensual route in Europe (and Japan). This puts into doubt any easy conclusion that antagonistic institutions provide for less environmental protection than more consensual ones. One possible explanation for the different research results is that this study was able to look at a longer time period. While their studies were published in the mid-1980s, this study was able to incorporate the ten years after 1986 as well. As environmental policy processes are usually played out over the long term, perhaps the full effects of institutions on environmental protection can only be discovered in the long run. If this reasoning had some truth to it, it would make this analysis somewhat more revealing. In this respect, it is significant that Richard Andrews has recently concluded that water pollution in the United States has probably grown slightly worse over time.[93]

When empirical evidence is not sufficient, issues can also be tackled deductively. A rationale can be set up suggesting that consensual institutions benefit environmental protection more than divisive ones, as follows. Ecological issues are truly *cross-boundary*, crossing both territorial and scientific borders, and also cutting across different segments of society as both their causes and solutions typically lie in a variety of social processes. It can be argued that ecological issues are so complex that their resolution needs the cooperation of all involved organizations. Each of these organizations has unique skills and knowledge. Firms have detailed knowledge of their cost structure, and are well-positioned to develop new technologies and find efficient, practical solutions to environmental problems. Environmental groups are useful 'watchdogs'. They tend to perceive ecological problems before other organizations do. Government agencies can be useful by acting as the 'objective', neutral arbitrator between the contradictory opinions of environmental groups and firms. They can also exert pressure on firms that stubbornly resist the implementation of environmental agreements, thus ensuring a level playing field. Furthermore, government agencies can set priorities and overview implementation, as well as coordinate different environmental measures. Under institutions that do not pit these various organizations against each other, the positive contributions that they all may have to offer to environmental protection are allowed to surface. Under adversarial institutions, everyone is busy discrediting the claims of everyone else, thereby reducing the contributions that each could make. This line of reasoning clearly favours more cooperative institutions for environmental protection. But the final verdict, of course, remains open.

Conclusion

In the present chapter an attempt was made to illustrate that institutions also matter in transboundary relations. It was argued that those institutions matter that regulate the interactions between organizations with different cultural dispositions. Certain specific institutions have the tendency to polarize the views and perceptions of adherents to different rationales. These adversarial institutions are often the outcome of a desire to limit the power of central government. Other, more consensual institutions tend to diminish differences between alternative cultural views. These are often institutions that induce people and organizations to accept the authority of central government. Public policy processes within a region are affected by the specific mix of adversarial and consensual institutions that exist in the area.

At least one question, it seems, remains to be answered. Do the results of this chapter negate the findings of previous chapters? In Chapters 4 and 5 it was illustrated that individualistic organizations can make important contributions to the solution of transboundary environmental problems, largely because of their inventive and pragmatic decision-making and because their behaviour is not greatly affected by the existence of state borders. Is this idea not contradicted by the lack of willingness of Great Lakes firms to invest in water protection, unless legally obliged to do so? After all, it has been argued that the views and actions of these firms could be categorized as individualistic, compared to the perceptions and deeds of other organizations in the Great Lakes watershed. The answer to this question seems simply to be 'no'. For several reasons the findings are not contradictory. In the first place, there is some evidence of a few, small voluntary protection measures taken by Great Lakes firms.[94] Admittedly, these appear to be much more limited than the investments made by firms in the Rhine basin. In addition, the policies and institutions set up with regard to environmental protection of the Great Lakes have simply destroyed any desire that firms might have had for taking environmental protection measures ahead of legislation. This was the thrust of the whole argument in this chapter. With regulations in place that were perceived as too rigid, impossibly strict as well as unfair, there never was any opportunity for extensive voluntary measures.

A second, and related, objection can be formulated thus. Do the results of this chapter signify that the institutional settings in which

environmental politics take place are always more important determinants of actors' behaviour than the cultures these actors adhere
to? In other words, why use cultures as explanatory factors at all
(as in Chapters 4 and 5)? At least two reasons can be given for
why cultures should be kept in. To start with, even if it were true
that institutional settings had more explanatory power than the
cultures that actors adhere to, these cultures could still explain the
decisions and moves of actors operating *within the same institutional
context*. This is essentially the course taken in Chapters 4 and 5
where it was shown that various organizations have made hugely
different contributions to the clean-up of the Rhine. The explanation for this rested to a large degree on the different ways of life
adhered to by these various organizations. Institutions could not
have served as an explanatory factor here, as the organizations all
operated in (more or less) the same institutional setting. Which
explanatory factors can and should be invoked depends on the research question.

Furthermore, it is our belief (but this is no more than a belief)
that institutions only matter in research comparing environmental
politics in regions where institutional settings diverge significantly
– such as Western Europe and the United States. Even then, one
would argue, only the *interplay* between institutional differences and
cultural settings (the ways in which alternative institutions structure the relations between adherents of cultures in different modes)
provides a full explanation. Of course then the problem becomes:
how much institutional divergence constitutes a 'significant' divergence? Put somewhat differently: when are institutional settings
different enough to become powerful explanatory factors in addition to cultures? The next, final chapter of this book will suggest
an answer to that question.

7
Conclusion: Cultures and Institutions Both Matter in Transboundary Relations

Cultures matter

This book has staked the claim that cultures matter in transboundary relations. More precisely, it has been argued that the four ways of life (or cultures) that are distinguished in grid-group theory (egalitarianism, fatalism, individualism and hierarchy) are useful ideal types with which to describe and explain the thoughts and actions of actors involved in transboundary regimes. Different regime actors have alternative perceptions of how transboundary relations should be organized, have divergent reactions to international anarchy, have varying perceptions of transboundary issues and disagree on how transboundary issues should be tackled.

Poststructuralists and (most) constructivists have suggested that the ways in which actors constitute the international system are infinite. According to these writers, the most we can hope for is to be able to trace the particular modes of thought and action that international actors have engaged in at specific points in time and space. Perhaps this is indeed all we can hope for – many arguments could be made to this effect. The failure of the more 'totalizing' paradigms (neorealism and neoliberalism for instance) also seems to lend support to this position. But perhaps there are grounds for hoping for something more. It may not be necessary to choose between 'historical' and 'scientific' approaches to international relations. There need be no obligation to opt either for perceiving only one rationality guiding international actors, or for claiming that an unending number of rationalities have existed in the international realm. Grid-group theorists call themselves 'constrained relativists'. They acknowledge that social reality is 'co-constituted' by actors and social relations. But they also postulate that the number of ways in which social reality can be constructed need not be unlimited. Yes, there will always be the specifics of time and space. But underlying these specifics may be a limited set of basic ways (ideal types) according to which people think and organize themselves. To assume the opposite seems to place a very high premium on human creativity and communication.

This study has attempted to extend cultural theory's four ways of life to include preferences and perceptions regarding transboundary issues. It has been argued that decision-makers within an international issue-area tend to choose between four alternative sets of international principles, norms, rules and procedures. Each of these divergent sets of volitions encompasses a specific: (a) view of how international governance should be organized; (b) conception of peace; (c) belief in the feasibility of transboundary cooperation; (d) understanding of the essence of international issues; and (e) perception of how transboundary issues should be managed. Sometimes, actors combine several of these rationalities, but often their actions and conceptions can be adequately described by a single international way of life, especially when compared to the deeds and thoughts of other decision-makers. Each actor strives to get their own views and preferences institutionalized within the international issue-area.

These theoretical notions have been illustrated in the empirical part of this study, especially in Chapters 4 and 5. It has been shown that groups of organizations involved in the pollution and restoration of the Rhine and the Great Lakes have tended to follow a variety of rationalities that strongly resemble the ways of life set out by cultural theory. Most of these groups of organizations have tended to adhere to a single culture, although some have combined cultures. An example of the latter has been the environmental organization World Wide Fund for Nature – Germany, whose discourse has included both hierarchical and egalitarian notions. In Chapters 4 and 5 it was argued that international policy decisions regarding the protection of the Rhine, as well as investments in water protection by Rhine firms, can be understood in terms of the cultures to which actors adhere.

One cultural phenomenon has been particularly striking. Actors who have tended to think and organize along hierarchical lines have found it more difficult to develop extensive cooperation in the anarchic international realm than have more individualistically oriented actors. This was exemplified in the Rhine case. Until 1987, national delegations, favouring hierarchical principles, norms, rules and procedures, could not agree on any intergovernmental policies for the protection of the Rhine, despite: (a) frantic efforts to do so; (b) the simultaneous development of extensive domestic water protection policies; and (c) massive voluntary investments in water protection by more individualistic private actors. In 1987, it was a project team of individualistically oriented McKinsey consultants that formulated the principles and norms that launched the exemplary present-day international cooperation on the protection of the Rhine. This was the basic message of Chapter 4. In the subsequent Chapter 5, it was shown that the individualistic principles, norms and programmes that have been followed by the large companies along the Rhine can also be credited for a significant part of the clean-up of the river. Of course, it could be argued on the basis of Chapter 6 that an adherence to individualism is not a waterproof guarantee of willingness to cooperate transnationally. In that chapter it was made clear that the companies located in the Great Lakes watershed have strictly adhered to individualistic principles, yet have been quite unwilling to invest in the environmental protection of the water basin. However, it also became clear in Chapter 6 that this behaviour has not been rooted in any reluctance to cooperate internationally *per se*. The Great Lakes firms have not been willing to invest in water protection mainly because they have sincerely believed that the governmental aims for the watershed have been unfair to them: too strict, based on 'bad science', not justified by proper cost–benefit analyses and implemented too harshly. Any unwillingness to contribute to a transboundary issue may not have had not much to do with it.

   This is certainly not a claim that the usefulness of grid-group theory for describing and explaining the decisions and mindsets of international actors has been proven or shown in any conclusive way. Given the limits of only two empirical cases, it is hoped that the usefulness of this approach has been suggested.

Institutions matter as well

Chapter 6 illustrated that institutions matter as well in transboundary relations. Institutions are self-sustaining modes of thought and practice. They are different from cultures in being background variables. Institutions matter as they structure the relations between organizations following alternative ways of life. Some institutional settings simply and crudely exclude actors adhering to a particular culture from decision-making processes while others have more subtle effects. These sets of institutions influence transboundary relations either by bringing closer together organizations that follow alternative ways of life or, in contrast, by dividing them. The former sets of institutions may be called 'consensual', the latter 'polarizing' or 'adversarial'. In a consensual institutional setting, actors will respect and understand each other's viewpoints more. Their ways of life will also include more overlapping elements than in adversarial institutional settings. Not too surprisingly, cooperation between actors following different ways of life will be much easier to achieve in a consensual institutional setting. Under adversarial institutions, followers of a certain way of life will deny the legitimacy of the viewpoints and actions of the adherents to other ways of life. Under such institutions, actors will only value their own particular viewpoints, and will fight as hard as they can to impose these viewpoints on others.

From the above it also follows which institutions are important: those institutions that regulate the relations between stakeholders in an issue-area. In environmental issue-areas, the main stakeholders usually include governmental agencies, the judiciary, parliament, environmental groups and firms. All institutions that impinge on the relations between these actors are therefore relevant. In Chapter 6, it was shown that the institutions shaping the relations between the executive, legislature, judiciary, environmental movements and corporations have been much more adversarial in North America than in the countries through which the Rhine flows. As a consequence, the ways of life adhered to by stakeholders have tended to be more 'extreme' in the Great Lakes basin than in the Rhine valley. In particular, Great Lakes companies have disagreed much more vehemently with governmental agencies and environmental groups over both the causes and effects and the extent of water pollution than Rhine firms have. This at least partly explains why Great Lakes firms have made fewer voluntary and obligatory investments in water protection than their counterparts in the Rhine catchment area.

In sum, the theoretical and empirical analysis developed in these pages suggests a second set of variables that affect transboundary environmental issues: the institutions that shape and regulate the relations between stakeholders adhering to alternative ways of life.

When cultures and institutions matter together

Generally speaking, cultures and institutions matter together. The latter structure the clashes between actors following different cultures, and the outcomes of these cultural clashes reshape the institutions in which the actors operate. However, in Chapters 4 and 5 cultures were used only as explanatory variables and it was chosen to 'neglect' institutions. This raises the question: when can one employ only cultures as explanatory factors, and when does one need to base one's explanation on both cultures and institutions? This is, of course, a very general and difficult question to which the response can only be tentative.

First, everything depends on the research question that informs the empirical analysis. In Chapter 5, the contributions that business corporations have made to the protection of the Rhine were compared with the contributions made by other groups of organizations (such as farmers and government agencies) to the revival of the Rhine basin. By and large, these various types of organization have operated in the same institutional context. As a consequence, in Chapter 5 the institutional context could not explain the differences in cooperativeness between the various groups of actors in the Rhine area. By contrast, in Chapter 6, the clean-up of the Great Lakes was compared with the restoration of the Rhine. Industrial polluters in these two areas have faced quite different institutional settings. In Chapter 6 these institutional differences were more important than cultural differences in explaining why polluters in the Rhine valley have invested more extensively in water protection than polluters in the Great Lakes basin.

But this brings up a further question: what are 'different' institutional settings? More precisely: when are institutional settings different enough to become important explanatory factors in addition to cultures? The following answer is suggested: institutional settings are sufficiently different from each other (to become explanatory variables) when they have been based on alternative combinations of the four cultures spelled out by grid-group theory. Let us clarify this idea. As explained in Chapters 2 and 3, each culture incorporates preferences regarding the ways in which social relations should be organized. These preferences also pertain to the relations between the executive, legislature, judiciary, electorate, citizens' groups and business firms. Within societies, people with different cultural biases will strive to get their viewpoints accepted and institutionalized. At particular, 'historic' points in time, these cultural struggles will (at least for some time) be settled. An obvious example of such a historic moment is the conference at which the US Constitution was drafted. Usually, the institutions that result from such historic cultural clashes will be a mix of the plans put forward by followers of the various ways of life.[1] Yet, the weight that will be accorded to each way of life in the resulting institutional setting will differ from society to society (and from era to era). If the institutional settings have sprung from alternative combinations of the four ways of life, then it can be expected that these institutional settings are sufficiently different from each other to become important explanatory variables in their own right. For instance, two institutional settings are 'sufficiently different' when one of them is dominated by two of the ways of life spelled out by grid-group theory, while the other setting represents two other ways of life, or three ways of life. In such cases, it is likely that institutional differences will have to be included as explanatory factors alongside cultural differences.

Chapter 6 provides a clear example. All the authors consulted on 'American exceptionalism' agree on one thing: American institutions are (almost) unique, as they are based on an abhorrence, and rejection, of hierarchy. Fear of strong, centralized power has been at the heart of the American system of checks and balances. Aaron Wildavsky, Richard Ellis and Daniel Elazar have actually shown that American exceptionalism has consisted of a (sometimes fragile, but enduring) coalition between an egalitarian and an individualistic political culture. The hierarchical way of life has been conspicuously absent (or, at least, under-represented) in the United States.[2] In comparison, the institutions prevalent in the Rhine countries have included many more hierarchical elements, in addition to egalitarian and individualistic principles and rules. In other words, American institutions have been permeated by only two ways of life, whereas West European institutions have usually been based on three ways of life. As a result, environmental policy processes have been quite different in both areas. In the United States relations between stakeholders in environmental issue-areas have been much more adversarial than in the Rhine countries. In Chapter 6, it was shown how these institutional differences could account for the greater degree of water protection that has been achieved in the Rhine valley, as compared to the water protection in the Great Lakes basin.

Robert Putnam's excellent work on democracy in Italy offers another example.[3] Putnam shows that the regional governments in the southern provinces of Italy have functioned much less satisfactorily than the regional governments in other Italian provinces. He explains these differences in terms of the lower levels of interpersonal trust that have existed in Southern Italy. In terms of cultural theory, Putnam is arguing that fatalism (as a way of life) is much more widespread and deep-rooted in the southern parts of Italy than in the central and northern parts of that country. As a consequence, policy outcomes in the former parts differ significantly from policy results in the latter areas.

These two empirical examples illuminate the following point: if the institutional settings of issue-areas represent alternative mixes of ways of life, then institutional differences become powerful explanatory variables in conjunction with cultures.

How to explain a paradox

The principal aim of this study has been to explore whether, and how, the cultural theory of the Douglas school can be usefully employed in the study of international relations. If anything, this cultural theory sometimes seems somewhat better equipped to provide explanations for paradoxical developments in politics, and several such paradoxes appeared in the empirical research presented in this book. This is not referring to the puzzles with which each empirical chapter started which were mystifying enough, but paradoxes which are more astonishing than the puzzles. Three developments in particular contradicted both common sense and received wisdom in the study of international relations.

First, there was the case of large industrial corporations in the Rhine basin protesting against the adoption of governmentally imposed effluent limits, while at the same time taking voluntary water protection measures that were going beyond these effluent standards. Cultural theory is able to provide a partial explanation for this strange case. In Chapter 5 it was shown that Rhine companies tended to follow individualistic principles and procedures. A central part of the individualistic culture is the desire to limit the size of government, at both the national and international level. According to individualists, the economy and society should not be overly regulated. This partly explains why Rhine firms protested against

the adoption of governmental effluent limits. The Rhine corporations were not financially hurt by proposed effluent limits and
technological standards, but the proposals did not accord with
their views on how the international political economy should be
shaped. Hence, it was rational for them to protest against governmental measures and policies, while at the same time going beyond
the standards set by the governments.

The second paradox that has been brought up in this research
concerns the agricultural sector. In both the United States and the
Rhine countries, policy-makers are convinced that many farmers in
their countries would be able to reap the same harvest if they cut
their use of fertilizers by half. Experiments by agricultural experts
are said to have proven this. West European farmers have been
extremely reluctant to follow this advice. In other words, agriculturists have on the whole not been willing to embark on a course
that would probably keep revenues steady and cut costs, as well as
spare the environment. Given the dire financial straits in which
many European farmers have found themselves in recent times, this
is paradoxical. Cultural theory would explain this paradox in terms
of the increasingly fatalistic environment, and corresponding rationality, of West European farmers. These farmers have become more
and more subject to environmental and other regulations during
the last few years. They have been forced to report many aspects
of their activities to the authorities. Financial support from the EU
has been eroded. Agricultural interest organizations have split up.
Relations with the Ministries of Agriculture have become strained.
These developments all represent movements towards either lower
'group' (less solidarity) or higher 'grid' (more external regulation).
Hand in hand with these increasingly fatalistic social relations have
gone more fatalistic attitudes among farmers. A fatalistic cultural
bias includes a large dose of pessimism as well as a short-term horizon.
It also encourages people to avoid risks at all costs, even when this
means forgoing opportunities to make money or improve life in
other ways. This is probably part of what has gone on in the agricultural sector of the European Union. With their backs against
the wall, a number of farmers have probably lost faith in expert
advice as well as willingness to try out ideas that deviate from established patterns.

The last paradox has received ample attention in this study. It
consists of the fruitless efforts of government agencies to agree on
international measures for the protection of the Rhine at a time

when the same agencies adopted and implemented extensive domestic water protection policies. This situation lasted from the early 1970s until 1987. It is explicable in terms of the hierarchical principles, norms, rules and procedures that were followed by these officials during 1970–86. In each of the Rhine countries elaborate water protection programmes were in place, containing effluent limits, water quality standards, technology prescriptions and pollution taxes. These extensive top-down approaches were not compatible with the intent of public officials to develop a similar sophisticated, hierarchical approach at the international level. Also, as hierarchists, the governmental actors felt that they had to follow the painstaking procedures of official international treaty-making. Lastly, state officials were inclined to think in terms of 'us' versus 'them' and to distrust representatives of other states – again as hierarchists are wont to do.

But what is a paradox anyway? The *Longman Dictionary of the English Language* used in writing this book states it is 'a tenet contrary to received opinion'. So what is seen as paradoxical partly depends on the implicit and explicit theories that inform our thinking. The three developments described above are only paradoxes if one has a narrow conception of actors as being solely motivated by the need to make money and feel secure. By providing a more elaborate picture of human motivation, grid-group theory is able to complement and enrich received opinion and established theories.

This applies with equal force (or so we would like to suggest) in the study of international relations.

Appendix A
Interviews with Stakeholders in the Environmental Protection of the Rhine

(* Signifies an interview over the telephone.)

Governmental organizations

The Netherlands

Ministry of Transport, Public Works and Watermanagement, The Hague, 14 March 1996; and Haarlem, 4 April 1996.
Ministry of the Environment, The Hague, 14 March 1996.
Ministry of Foreign Affairs, The Hague, 25 March 1996; The Hague, 4 April 1996; and Amsterdam, 8 April 1996.
Ministry of Agriculture, Nature and Fishery, The Hague, 11 April 1996.
Port of Rotterdam, Rotterdam, 12 April 1996.
Dr Pieter Winsemius, former Minister of the Environment, presently director of McKinsey Consultancy Amsterdam, Amsterdam, 5 July 1996.
Institute for Inland Water Management and Wastewater Treatment (RIZA), Lelystad, 1 August 1996; and Lelystad, 24 January 1997.*
Ms Neelie Kroes, former Minister of Transport, Public Works and Water Management, Nijenrode, 20 August 1996.

Federal Republic of Germany

Ministerium für Umwelt, Energie, Jugend, Familie und Gesundheit des Landes Hessen, Wiesbaden, 11 July 1996.
Ministerium für Umwelt des Landes Nordrhein-Westfalen, Düsseldorf, 23 July 1996.*
Ministerium für Umwelt des Landes Rheinland-Pfalz, Mainz, 24 July 1996.
Bundesministerium für Umwelt, Bonn, 5 August 1996.
Ministerium für Umwelt und Verkehr des Landes Baden-Württemberg, Stuttgart, 11 November 1996.
Municipality of Bonn, Bonn, 11 March 1997.*
Municipality of Mannheim, Mannheim, 7 April 1997 (written answers).
Municipality of Frankfurt, Frankfurt, 11 April 1997.*

Luxembourg

Administration de l'Environnement, Luxembourg, 12 July 1996.

France

Agence de l'Eau Rhin–Meuse, Metz, 13 November 1996.
Ministère de l'Agriculture, de la Pêche et de l'Alimentation, Paris, 19 November 1996.
Ministère de l'Environnement, Paris, 19 November 1996.

Switzerland

Bundesamt für Umwelt, Wald und Landschaft, Bern, 14 November 1996.

International organizations

European Commission, Directorate General XI, Brussels, 2 April 1996; and Brussels 3 April 1996.
European Parliament, Brussels, 2 April 1996.
International Commission for the Protection of the Rhine against Pollution (ICPR), Koblenz, 8 July 1996; Delft, 30 July 1996; Koblenz, 5 August 1996; and The Hague, 8 November 1996.
Central Commission for the Navigation of the Rhine, Strasbourg, 15 November 1996.

Business

European Chemical Industry Council (CEFIC), Brussels, 3 March 1996.
Shell Pernis BV, Rotterdam, 18 March 1996.
Internationale Arbeitsgemeinschaft der Wasserwerke im Rheineinzugsgebiet (RIWA), Amsterdam, 5 April 1996.
Dr Pieter Winsemius, former Minister of the Environment, presently director of McKinsey Consultancy Amsterdam, Amsterdam, 5 July 1996.
Verband der Chemischen Industrie (VCI), Frankfurt, 10 July 1996.
Hoechst AG, Frankfurt, 11 July 1996.
BASF AG, Ludwigshaven, 24 July 1996.
Bayer AG, Leverkussen, 1 August 1996.*
Mines de Potasse d'Alsace, Mulhouse, 12 November 1996.
Sandoz Pharma AG, Basel 14 November 1996.
Ciba-Geigy AG, Basel, 14 November 1996.

Agriculture

Landbouwschap, The Hague, 25 March 1996.
Deutscher Bauernverband, Bonn, 10 November 1996.
Fédération Nationale des Syndicats d'Exploitants Agricoles, Paris, 17 March 1997.*

Environmental organizations

Stichting Reinwater, Amsterdam 20 March 1996; Amsterdam, 26 March 1996;* and Amsterdam, 12 April 1996.
World Wide Fund for Nature – Germany, Rastatt, 11 July 1996.

Naturschutzbund Deutschland, Kranenburg, 12 July 1996.
Bundesverband Bürgerinitiativen Umweltschutz, Freiburg, 11 November 1996.
Bund für Naturschutz Baselland, Basel, 15 November 1996.

Total number of interviews: 54
Total number of interviewees: 58

Appendix B
Interviews with Stakeholders in the Environmental Protection of the Great Lakes

(* Signifies an interview over the telephone.)

US governmental and tribal organizations

New York Department of Environmental Conservation, Albany, New York, 27 May 1997.*
Department of Agriculture, Natural Resources Conservation Service, Madison, Wisconsin, 29 May 1997.
Great Lakes Environmental Research Laboratory, Department of Commerce, Ann Arbor, Michigan, 3 June 1997.
Great Lakes Commission, Ann Arbor, Michigan, 6 June 1997.
Environmental Protection Agency, Great Lakes Regional Office, Chicago, Illinois, 6 June 1997.
Federal Reserve Bank of Chicago, Chicago, Illinois, 20 June 1997.
Environmental Protection Agency, Washington, DC, 23 June 1997.
Wisconsin Department of Natural Resources, Madison, Wisconsin, 7 July 1997.
Army Corps of Engineers, Chicago, Illinois, 7 July 1997.*
Fish and Wildlife Service, East Lansing, Michigan, 8 July 1997.*
State Department, Washington, DC, 9 July 1997.*
Michigan Department of Natural Resources, Great Lakes Office, East Lansing, Michigan, 15 July 1997.*
Environmental Protection Agency, Washington, DC, 16 July 1997.*
Great Lakes Indian Fish and Wildlife Commission, Adena, Wisconsin, 17 July 1997.*
Council of Great Lakes Governors, Chicago, Illinois, 25 July 1997.

Canadian governmental agencies

Ministère de l'Environnement et de la Faune, Gouvernement du Québec, Quebec, 28 May 1997.*
Agriculture Canada, Guelph, Ontario, 11 June 1997.
Ontario Minister of Environment and Energy, Guelph, Ontario, 11 June 1997.
Environment Canada, Downsview, Ontario, 12 June 1997.
Ontario Ministry of Environment and Energy, Toronto, Ontario, 24 July 1997.*

Canadian Center for Pollution Prevention, Industry Canada, Sarnia, Ontario, 10 July 1997.*

International organizations

Great Lakes Fishery Commission, Ann Arbor, Michigan, 3 June 1997.
International Joint Commission, Great Lakes regional office, Windsor, Ontario, 5 June 1997.
International Joint Commission, US section, Washington, DC, 24 June 1997.
International Joint Commission, Canadian section, Ottawa, 17 July 1997.*

Business

Council of Great Lakes Industries, Ann Arbor, Michigan, 4 June 1997.
Chemical Manufacturers Association, Arlington, Virginia, 24 June 1997.
Chlorine Chemistry Council, Arlington, Virginia, 24 June 1997.
American Forest and Paper Association, Washington, DC, 25 June 1997.
LTV Steel, Cleveland, Ohio, 8 July 1997.*
Canadian Chlorine Coordinating Committee, Ottawa, 17 July 1997.*
Inland Steel, East Chicago, Indiana, 18 July 1997.*
Xerox Company, Buffalo, New York, 21 July 1997.*
Wisconsin Manufacturers and Commerce, Madison, Wisconsin, 21 July 1997.
Ford Motor Company, Dearborn, Michigan, 30 July 1997.*
Eastman Kodak Company, Rochester, New York, 4 August 1997.*
Great Lakes Water Quality Coalition, Chicago, Illinois, 4 August 1997.*
Canadian Vehicle Manufacturers' Association, Toronto, 5 August 1997.*

Environmental organizations

Nature Conservancy, Chicago, Illinois, 2 June 1997.
National Wildlife Federation, Ann Arbor, Michigan, 4 June 1997.
Pollution Probe, Toronto, Ontario, 13 June 1997.
Sierra Club – Great Lakes Office, North Henry Street, Madison, Wisconsin, 19 June 1997.
Greenpeace, Great Lakes Office, Chicago, Illinois, 20 June 1997.
Environmental Defense Fund, Washington, DC, 27 June 1997.
Great Lakes United, Buffalo, New York, 9 July 1997.*
Great Lakes Tomorrow, Toronto, Ontario, 10 July 1997.*
Canadian Institute for Environmental Law and Policy (CIELAP), Toronto, Ontario, 14 July 1997.*

Total number of interviews: 47
Total number of interviewees: 51

Notes

Foreword

1 Mary Douglas and Aaron Wildavsky, *Risk and Culture: An Essay on the Selection of Technical and Environmental Dangers* (Berkeley, CA: University of California Press, 1982).

Chapter 1   Introduction

1 Together these three approaches formed the 'inter-paradigm debate'. An introduction is M. H. Banks, 'The Inter-Paradigm Debate', in M. Light and A. J. R. Groom (eds), *International Relations: A Handbook of Current Theory* (London: Frances Pinter, 1985).

2 The literature criticizing traditional IR approaches and calling for new paradigms and research programmes is too extensive to be fully covered here. Some efforts are M. Hofmann, 'Critical Theory and the Inter-Paradigm Debate', *Millennium: Journal of International Studies*, vol. 16, 1987, pp. 231–49; A. L. Wendt, 'The Agent–Structure Problem in International Relations Theory', *International Organization*, vol. 41, 1987, pp. 335–70; and his 'Anarchy Is What States Make of It: The Social Construction of Power Politics', *International Organization*, vol. 46, 1992, pp. 391–425; *Women and International Relations*, special issue of *Millennium: Journal of International Studies*, vol. 17, 1988; *Culture in International Relations*, special issue of *Millennium: Journal of International Studies*, vol. 22, 1993; N. G. Onuf, *World of Our Making: Rules and Rule in Social Theory and International Relations* (Columbia, SC: University of South Carolina Press, 1989); Y. Lapid, 'The Third Debate: On the Prospects of International Theory in a Post-Positivist Era', *International Studies Quarterly*, vol. 33, 1989, pp. 235–54; J. Der Derian and M. J. Shapiro (eds), *International/Intertextual Relations: Postmodern Readings of World Politics* (New York: Lexington Books, 1989); J. N. Rosenau, *Turbulence in World Politics: A Theory of Change and Continuity* (New York: Harvester-Wheatsheaf, 1990); Y. H. Ferguson and R. W. Mansbach, 'Between Celebration and Despair: Constructive Suggestions for Future International Theory', *International Studies Quarterly*, vol. 35, 1991, pp. 363–86; A. Linklater, 'The Question of the Next Stage in International Relations Theory: A Critical-Theoretical Point of View', *Millennium: Journal of International Studies*, vol. 21, 1992, pp. 77–98; R. B. J. Walker, *Inside/Outside: International Relations as Political Theory* (Cambridge: Cambridge University Press, 1993); F. Halliday, *Rethinking International Relations* (London: Macmillan, 1994); F. Kratochwil and Y. Lapid (eds), *The Return of Culture and Identity in International Relations Theory* (Boulder, CO: Lynne Rienner Press, 1995). The admonition to use our 'international imagination' is

from J. Rosenberg, 'The International Imagination', *Millennium: Journal of International Studies*, vol. 23, 1994, pp. 85–108.

3   T. J. Biersteker, 'Critical Reflections on Post-Positivism in International Relations', *International Studies Quarterly*, vol. 33, 1989, p. 266.

4   R. O. Keohane, 'International Institutions: Two Approaches', in his *International Institutions and State Power* (Boulder, CO: Westview Press, 1989), pp. 170–4.

5   The lack of theory in newly established approaches to international relations is also lamented in M. Finnemore, *National Interests in International Society* (Ithaca, NY: Cornell University Press, 1996), p. 130; J. T. Checkel, 'The Constructivist Turn in International Relations Theory', *World Politics*, vol. 50, 1998, pp. 324–48; and Y. H. Ferguson, 'Looking Backwards at Contemporary Polities', in D. V. F. Jacquin-Berdal, A. Oros and M. Verweij (eds), *Culture in World Politics* (London: Macmillan, 1998).

6   See in particular R. K. Ashley and R. B. J. Walker, 'Speaking the Language of Exile: Dissident Thought in International Studies', *International Studies Quarterly*, vol. 34, 1990, p. 266. Also F. Kratochwil, *Theory and Political Practice: Reflections on Theory-Building in International Relations*, paper presented at the European University Institute (San Domenico di Fiesole, Italy, December 1997).

7   Empirical research based on poststructuralist ideas includes D. Campbell, *Writing Security: United States Foreign Policy and the Politics of Identity* (Minneapolis, MN: University of Minnesota Press, 1992); I. B. Neumann and J. M. Welsh, 'The Other in European Self-Definition', *Review of International Studies*, vol. 17, 1991, pp. 327–48; J. Der Derian, *Anti-Diplomacy: Spies, Terror, Speed, and War* (New York: Blackwell, 1992); R. L. Doty, 'Foreign Policy as Social Construction: A Post-Positivist Analysis of U.S. Counterinsurgency Policy in the Philippines', *International Studies Quarterly*, vol. 37, 1993, pp. 297–320; and K. T. Litfin, *Ozone Discourses: Science and Politics in Global Environmental Cooperation* (New York: Columbia University Press, 1994).

8   R. L. Doty, 'The Bounds of "Race" in International Relations', *Millennium: Journal of International Studies*, vol. 22, 1993, pp. 443–61.

9   E. A. Adler, 'Seizing the Middle Ground: Constructivism in World Politics', *European Journal of International Relations*, vol. 3, 1997, pp. 319–63.

10 Examples of empirical analyses based on constructivist assumptions are R. D. Lipschutz and K. Conca (eds), *The State and Social Power in Global Environmental Politics* (New York: Columbia University Press, 1993); P. J. Katzenstein (ed.), *The Culture of National Security* (New York: Columbia University Press, 1996); T. J. Biersteker and C. Weber (eds), *State Sovereignty as Social Construct* (Cambridge: Cambridge University Press, 1996); M. Finnemore, *National Interests in International Society* (Ithaca, NY: Cornell University Press, 1996).

11 J. T. Checkel, 'The Constructivist Turn in International Relations Theory', *World Politics*, vol. 50, 1998, p. 338.

12 Respectively, M. Barnett, *Dialogues in Arab Politics* (New York: Columbia University Press, 1998); E. B. Haas, *Nationalism, Liberalism and Progress* (Ithaca, NY: Cornell University Press, 1997); and A. E. Wendt, *Social Theory in International Politics* (Cambridge: Cambridge University Press, 1999).

13   S. Lieberson, 'Einstein, Renoir, and Greeley: Some Thoughts about Evidence in Sociology', *American Sociological Review*, vol. 56, 1992, pp. 1–15.
14   G. King, R. O. Keohane and Sydney Verba, *Designing Social Inquiry: Scientific Inference in Qualitative Research* (Princeton, NJ: Princeton University Press, 1994).
15   On this research strategy, see S. Bartolini, 'On Time and Comparative Research', *Journal of Theoretical Politics*, vol. 5, 1993, pp. 131–67.
16   The counterfactual research strategy is discussed in J. D. Fearon, 'Counterfactuals and Hypothesis Testing in Political Science', *World Politics*, vol. 43, 1991, pp. 169–95. In this essay, Fearon actually argues that any non-experimental causal reasoning necessarily uses elements of the counterfactual approach.
17   A. Lijphart, 'Comparative Politics and the Comparative Method', *American Political Science Review*, vol. 65, 1971, pp. 682–93; D. Collier, 'The Comparative Method', in A. W. Finifter (ed.), *Political Science: The State of the Discipline II* (Washington, DC: American Political Science Association, 1993). This particular form of cross-spatial research is usually called the 'comparative method', or 'John Stuart Mill's method of difference', or 'most similar systems design'. See King, Keohane and Verba, op. cit. in note 14, p. 168.
18   For instance, Gary King, Robert Keohane and Sydney Verba have emphasized the methodological difficulties of using concepts such as norms, perceptions and meanings in causal reasoning. See King, Keohane and Verba, op. cit. in note 14, pp. 109–12 and 191–3. Sociologists Robert Wuthnow, James Davidson Hunter, Albert Bergesen and Edith Kurzweil have also claimed that the use of such concepts is problematic in rigorous empirical research. See their *Cultural Analysis: The Work of Peter L. Berger, Mary Douglas, Jürgen Habermas and Michel Foucault* (London: Routledge, 1984), p. 255.
19   H. M. Kritzer, 'The Data Puzzle: The Nature of Interpretation in Quantitative Research', *American Journal of Political Research*, vol. 40, 1996, pp. 1–32.
20   A. S. Yee, 'The Causal Effects of Ideas on Policies', *International Organization*, vol. 50, 1996, pp. 69–108.
21   A. L. George, 'Case Studies and Theory Development: The Method of Structured, Focused Comparison', in P. G. Lauren (ed.), *Diplomacy: New Approaches in History* (New York, 1979); and A. L. George and T. J. McKeown, 'Case Studies and Theories of Organizational Decision-Making', in L. S. Sproull and P. D. Larkey, *Advances in Information Processing in Organization II* (Greenwich, CN: Jai Press, 1985).
22   I have not fully obeyed this injunction. In Chapters 4 and 5 I focus solely on the environmental protection of the Rhine, and do not analyse the protection of the Great Lakes.
23   See Appendices A and B.
24   Rosenau, op. cit. in note 2, p. 6.
25   J. Rosenberg, 'What's the Matter with Realism?', *Review of International Studies*, vol. 16, 1991, pp. 285–303.

Chapter 2   Grid-Group Theory and the Study of International Relations

1   Introductions to, and overviews of, the constructivist school are A. L. Wendt, 'Constructing International Politics', *International Security*, vol. 20, 1995, pp. 71–81; R. L. Jepperson, A. L. Wendt and P. J. Katzenstein, 'Norms, Identity, and Culture in National Security', in P. J. Katzenstein (ed.), *The Culture of National Security* (New York: Columbia University Press, 1996); E. A. Adler, 'Seizing the Middle Ground: Constructivism in World Politics', *European Journal of International Relations*, vol. 3, 1997, pp. 319–63; and J. T. Checkel, 'The Constructivist Turn in International Relations Theory', *World Politics*, vol. 50, 1998, pp. 324–48.

2   P. Rosenau, 'Once Again into the Fray: International Relations Confronts the Humanities', *Millennium: Journal of International Studies*, vol. 19, 1990, pp. 83–110.

3   C. Brown, '"Turtles All the Way Down": Anti-Foundationalism, Critical Theory and International Relations', *Millennium: Journal of International Studies*, vol. 23, 1994, pp. 213–36.

4   The following paper comes close to espousing this line of thought: F. Kratochwil, *Theory and Political Practice: Reflections on Theory-Building in International Relations*, paper presented at the European University Institute (San Domenico di Fiesole, Italy, December 1997).

5   This is the agent-structure problem. See A. L. Wendt, 'The Agent-Structure Problem in International Relations Theory', *International Organization*, vol. 41, 1987, pp. 335–70; 'Anarchy Is What States Make of It: The Social Construction of Power Politics', *International Organization*, vol. 46, 1992, pp. 391–425; and 'Collective Identity Formation and the International State', *American Political Science Review*, vol. 88, 1994, pp. 384–96; D. Dessler, 'What Is at Stake in the Agent-Structure Debate?', *International Organization*, vol. 43, 1989, pp. 441–73; and W. Carlsnaes, 'The Agent-Structure Problem in Foreign Policy Analysis', *International Studies Quarterly*, vol. 36, 1992, pp. 245–70.

6   P. J. Katzenstein, *Analyzing Change in International Politics*, Discussion Paper 90\10 (Cologne: Max-Planck-Institut für Gesellschaftsforschung, 1990); E. Adler, 'Cognitive Evolution: A Dynamic Approach for the Study of International Relations and Their Progress', in E. Adler and B. Crawford (eds), *Progress in Postwar International Relations* (New York: Columbia University Press, 1991); J. G. Ruggie, 'Territoriality and Beyond: Problematizing Modernity in International Relations', *International Organization*, vol. 47, 1993, pp. 139–74; B. Jahn, 'Globale Kulturkämpfe oder einheitliche Weltkultur? Zur Relevanz von Kultur in den Internationalen Beziehungen', *Zeitschrift für Internationale Beziehungen*, vol. 2, 1995, pp. 213–36.

7   On the need for reflexivity in the study of international relations see M. Neufeld, 'Reflexivity and International Relations Theory', *Millennium: Journal of International Studies*, vol. 22, 1993, pp. 53–76; on the need for reflexivity in the social sciences in general see P. Bourdieu, Pierre Wacquant and L. Wacquant, *An Invitation to Reflexive Sociology* (Cambridge: Polity Press, 1992).

8   M. Thompson, R. Ellis and A. Wildavsky, *Cultural Theory* (Boulder, CO: Westview Press, 1990).

9   M. Douglas, *Natural Symbols: Explorations in Cosmology* (London: Barrie & Rockliff, 1970); 'Cultural Bias', in her *In the Active Voice* (London: Routledge & Kegan Paul, 1982); M. Douglas and A. Wildavsky, *Risk and Culture: An Essay on the Selection of Technological and Environmental Dangers* (Berkeley, CA: University of California Press, 1982).

10 Other important theoretical contributions to this approach include A. B. Wildavsky, 'Choosing Preferences by Constructing Institutions: A Cultural Theory of Preference Formation', *American Political Science Review*, vol. 81, 1987, pp. 3–21; M. Schwarz and M. Thompson, *Divided We Stand: Redefining Politics, Technology and Social Choice* (Hemel Hempstead: Harvester-Wheatsheaf, 1990); M. Douglas, *Risk and Blame: Essays in Cultural Theory* (London: Routledge & Kegan Paul, 1992); *Thought Styles: Critical Essays on Good Taste* (London: Sage, 1996); Mary Douglas and Steven Ney, *Missing Persons* (Berkeley, CA: University of California Press, 1998). A collection of empirical applications is D. J. Coyle and R. J. Ellis (eds), *Politics, Policy, and Culture* (Boulder, CO: Westview Press, 1994).

11 For example, R. J. Ellis and M. Thompson (eds), *Culture Matters: Essays in Honor of Aaron Wildavsky* (Boulder, CO: Westview Press, 1997).

12 G. Mars, *Cheats at Work* (London: Allen & Unwin, 1982).

13 G. Engbersen, K. Schuyt, J. Timmer and F. van Waarden, *Cultures of Unemployment* (Boulder, CO: Westview Press, 1993).

14 R. J. Ellis and A. B. Wildavsky, *Dilemmas of Presidential Leadership: From Washington through Lincoln* (New Brunswick, NJ: Transaction Publishers, 1989); R. J. Ellis, *American Political Cultures* (Oxford: Oxford University Press, 1993); U. Edvarsen, 'A Cultural Approach to Understanding Modes of Transition to Democracy', *Journal of Theoretical Politics*, vol. 9, 1997, pp. 211–34; R. M. Coughlin and Ch. Lockhart, 'Grid-Group Theory and Political Ideology: A Consideration of Their Relative Strengths and Weaknesses for Explaining the Structure of Mass Belief Systems', *Journal of Theoretical Politics*, vol. 10, 1998, pp. 33–58; M. Thompson, G. Grendstad and P. Selle (eds), *Cultural Theory as Political Science* (London: Routledge, 1998).

15 A. B. Wildavsky, *The Beleaguered Presidency* (New Brunswick, NJ: Transaction Publishers, 1991); C. Hood, *The Art of the State: Culture, Rhetoric and Public Management* (Oxford: Clarendon Press, 1998); F. Hendriks, *Cultural Bias in Policy-Making: An Institutional Comparison of Cities* (Cheltenham: Edward Elgar, 1999).

16 M. E. A. Schmutzer, *Ingenium und Individuum: Eine sozialwissenschaftliche Theorie von Wissenschaft und Technik* (Berlin: Springer, 1994).

17 A. B. Wildavsky and K. Dake, 'Theories of Risk Perception: Who Fears What and Why?', *Daedalus*, vol. 119, 1991, pp. 41–60.

18 E. Sivan, 'The Enclave Culture', in M. M. Marty (ed.), *Fundamentalism Comprehended* (Chicago: Chicago University Press, 1995).

19 Exceptions are S. Rayner, 'A Cultural Perspective on the Structure and Implementation of Global Environmental Agreements', *Evaluation Review*, vol. 15, 1991, pp. 75–102; C. Jönsson, 'Cognitive Factors in Explaining Regime Dynamics', in V. Rittberger (ed.), *Regime Theory and*

International Relations (Oxford: Clarendon Press, 1993); and V. Ward, 'Towards an International Theory of State–Nonstate Relations: A Grid-Group Cultural Approach', in D. Jacquin-Berdal, A. Oros and M. Verweij (eds), *Culture in World Politics* (London/New York: Macmillan/St. Martin's Press, 1998).

20 The following description of the concepts of grid and group is taken from J. Gross and S. Rayner, *Measuring Culture: A Paradigm for the Analysis of Social Organization* (New York: Columbia University Press, 1985), pp. 5–7. The rest of their book is devoted to the issue of how to measure the concepts of grid and group in empirical research.

21 The grid-group distinction on which cultural theory is based was first developed by social anthropologist Mary Douglas. See especially her *Natural Symbols: Explorations in Cosmology* (London: Barrie & Rockliff, 1970). Douglas took her cue from the work of Emile Durkheim, whose concepts of 'integration' and 'regulation' closely resemble the grid and group dimensions. See E. Durkheim, *Suicide: A Study in Sociology* (New York: Free Press, 1966/1897).

22 In grid-group theory, egalitarianism has also been called 'the enclave culture' or 'sectarianism'.

23 Thompson, Ellis and Wildavsky, op. cit., in note 8, pp. 33–7.

24 Ibid., pp. 26–9.

25 Or, vice versa, one can try to predict people's political preferences on the basis of their views of and stances on specific policy issues. For such an attempt, see R. J. Ellis and F. Thompson, 'Culture and the Environment in the Pacific Northwest', *American Political Science Review*, vol. 91, 1997, pp. 885–97.

26 M. Douglas, *How Institutions Think* (London: Routledge & Kegan Paul, 1987).

27 In the early work of Douglas, ways of life are called 'cosmologies'. Lately, two other terms have also been used: 'rationalities' and 'social solidarities'. Please note that in this book I will use the terms 'ways of life' and 'cultures' interchangeably.

28 The best formulation of this is M. Thompson, G. Grendstad and P. Selle, 'Cultural Theory as Political Science', in Thompson, Grendstad and Selle, op. cit. in note 14.

29 F. Hendriks, 'Cars and Culture in Munich and Birmingham: The Case for Cultural Pluralism', in Coyle and Ellis, op. cit. in note 10; and Ch. Lockhart and G. Franzwa, 'Cultural Theory and the Problem of Moral Relativism', in Coyle and Ellis, op. cit. in note 10.

30 Ch. Lockhart, 'Political Culture and Political Change', in Ellis and Thompson, op. cit. in note 16.

31 D. Laitin, 'Political Culture and Political Preferences', *American Political Science Review*, vol. 82, 1988, pp. 589–97; J. Alexander and P. Smith, 'The Discourse of American Political Society: A New Proposal for Cultural Studies', *Theory and Society*, vol. 22, 1993, p. 153; V. Ostrom and E. Ostrom, 'Cultures: Frameworks, Theories, and Models', in Ellis and Thompson, op, cit. in note 11, pp. 80–2.

32 M. Weber, *Max Weber on Charisma and Institution Building* (Chicago: University of Chicago Press, 1968), especially part II.

33 Thompson, Ellis and Wildavsky, op. cit. in note 8, pp. 103–212.
34 Wildavsky, op. cit. in note 10, pp. 15–18.
35 J. Friedman, 'Accounting for Political Preferences: Cultural Theory v. Cultural History', *Critical Review*, vol. 5, 1991, pp. 332–7.
36 R. J. Ellis, 'The Case for Cultural Theory: Reply to Friedman', *Critical Review*, vol. 7, 1993, pp. 94–100.
37 W. Jann, 'Vier Kulturtypen die alles erklären? Kulturelle und institutionelle Ansätze der neueren amerikanischen Politikwissenschaft', *Politische Vierteljahresschrift*, vol. 27, 1986, p. 371; and P. Selle, 'Culture and the Study of Politics', *Scandinavian Political Studies*, vol. 14, 1991, pp. 105–7.
38 A. B. Wildavsky, 'What Other Theory Would Be Expected to Answer Such Profound Questions? A Reply to Per Selle's Critique of *Cultural Theory*', *Scandinavian Political Studies*, vol. 14, 1991, pp. 355–60.
39 J. Elster (ed.), *The Multiple Self* (Cambridge: Cambridge University Press, 1985).
40 E. Olli, 'Individual Level Rejection of Cultural Biases and Effects on Party Preference', in Thompson, Grendstad and Selle, op, cit. in note 14.
41 L. Sjöberg, 'Explaining Risk Perception: An Empirical Evaluation of Cultural Theory', *Risk Decision and Policy*, vol. 2, 1997, pp. 113–30.
42 D. J Coyle, 'The Theory That Would Be King', in Coyle and Ellis, op. cit. in note 10, pp. 231–3.
43 D. J. Coyle, '"This Land Is Your Land, This Land Is My Land": Cultural Conflict in Environmental Regulation and Land-Use Regulation', in Coyle and Ellis, op. cit. in note 10.
44 A. B. Wildavsky, *Culture and Social Theory* (New Brunswick, NJ: Transaction Publishers, 1998).
45 Examples of empirical analyses based on constructivist assumptions are R. D. Lipschutz and K. Conca (eds), *The State and Social Power in Global Environmental Politics* (New York: Columbia University Press, 1993); Katzenstein, op. cit. in note 1; T. J. Biersteker and C. Weber (eds), *State Sovereignty as Social Construct* (Cambridge: Cambridge University Press, 1996); M. Finnemore, *National Interests in International Society* (Ithaca, NY: Cornell University Press, 1996).
46 Finnemore, op. cit. in note 45, p. 130.
47 Checkel, op. cit. in note 1, p. 338; and Y. H. Ferguson, 'Looking Backwards at Contemporary Polities', in Jacquin-Berdal, Oros and Verweij, op. cit. in note 19.
48 P. Kowert and J. Legro, 'Norms, Identity, and Their Limits: A Theoretical Reprise', in Katzenstein, op. cit. in note 1, pp. 483–95.
49 G. Almond and S. Verba, *The Civic Culture* (Princeton, NJ: Princeton University Press, 1963).
50 G. Hofstede, *Culture's Consequences* (London: Sage, 1984).
51 M. H. Ross, *The Culture of Conflict* (New Haven, CN: Yale University Press, 1993).
52 Harry Eckstein agrees. See his 'Social science as Cultural Science, Rational Choice as Metaphysics', in Ellis and Thompson, op. cit. in note 11.
53 See O. Holsti, *The 'Operational Code' as an Approach to the Analysis of Belief Systems*. Final Report to the National Science Foundation, grant SOC 75–15368 (Durham, NC: Duke University, 1975); and S. Walker,

'The Evolution of Operational Code Analysis', *Political Psychology*, vol. 11, 1991, pp. 403–18.

54 The relevance of grid-group analysis to the study of revolutions and political violence is the subject of S. K. Chai and A. B. Wildavsky, 'Culture, Rationality, and Violence', in Coyle and Ellis, op. cit. in note 10.

55 S. Strange, *States and Markets* (London: Frances Pinter, 1988), pp. 1–6.

Chapter 3   Regimes, Institutions and Four Cultures

1 The two main publications on regime analysis are S. D. Krasner (ed.), *International Regimes* (Ithaca, NY: Cornell University Press, 1983); and V. Rittberger (ed.), *Regime Theory and International Relations* (Oxford: Clarendon Press, 1993). Overviews are M. Levy, O. Young and M. Zürn, 'The Study of International Regimes', *European Journal of International Relations*, vol. 1, 1993, pp. 267–330; and A. Hansenclever, P. Mayer and V. Rittberger, 'Interests, Power, Knowledge: The Study of International Regimes', *Mershon International Studies Review*, vol. 40, 1996, pp. 177–228.

2 S. Chan, 'Cultural and Linguistic Reductionisms and a New Historical Sociology for International Relations', *Millennium: Journal of International Studies*, vol. 22, 1993, pp. 425–7.

3 See E. B. Haas, 'Is There a Hole in the "Whole"?', *International Organization*, vol. 29, 1975, pp. 827–76; 'Why Collaborate? Issue-linkage and International Regimes', *World Politics*, vol. 32, 1980, pp. 357–405; and 'Words Can Hurt You; or, Who Said What to Whom about Regimes', *International Organization*, vol. 36, 1982, pp. 207–43; J. G. Ruggie, 'International Responses to Technology: Concepts and Trends', *International Organization*, vol. 29, 1975, pp. 557–83; 'On the Problem of "the Global Problematique": What Roles for International Organizations', *Alternatives*, vol. 5, 1980, pp. 517–50; 'International Regimes, Transactions and Change: Embedded Liberalism in the Postwar Economic Order', *International Organization*, vol. 36, 1982, pp. 379–415; O. A. Young, 'International Regimes: Problems of Concept Formation', *World Politics*, vol. 32, 1980, pp. 331–56.

4 S. D. Krasner, 'Structural Causes and Regime Consequences: Regimes as Intervening Variables', in Krasner, op, cit. in note 1.

5 F. Kratochwil and J. G. Ruggie, 'International Organization: A State of the Art on the Art of the State', *International Organization*, vol. 40, 1986, pp. 753–75. Other useful critiques of regime theory are S. Haggard and B. A. Simmons, 'Theories of International Regimes', *International Organization*, vol. 41, 1987, pp. 491–517; and G. Junne, 'Beyond Regime Theory', *Acta Politica*, vol. 27, 1992, pp. 9–28.

6 P. M. Haas, *Saving the Mediterranean: The Politics of International Environmental Protection* (New York: Columbia University Press, 1990); P. M. Haas (ed.), *Epistemic Communities and International Policy Coordination*, special issue of *International Organization*, vol. 46, 1992.

7 P. M. Haas, 'Introduction: Epistemic Communities and International Policy Coordination', *International Organization*, vol. 46, 1992, p. 3.

8 Christer Jönsson has also noted the relevance of cultural theory, and the ideas of Mary Douglas more generally, for regime analysis. See C.

Jönsson, 'Cognitive Factors in Explaining Regime Dynamics', in Rittberger, op. cit. in note 1.

9 J. F. Keeley, 'Toward a Foucauldian Analysis of International Regimes', *International Organization*, vol. 44, 1990, pp. 83–105.

10 S. Strange, '*Cave! Hic Dragones*: A Critique of Regime Analysis', in Krasner, op. cit. in note 1.

11 Levy, Young and Zürn, op. cit. in note 1, p. 274.

12 R. Jepperson, 'Institutions, Institutional Effects, and Institutionalism', in W. W. Powell and P. J. DiMaggio (eds), *The New Institutionalism in Organizational Analysis* (Chicago: University of Chicago Press, 1991), p. 145.

13 A. Zijderveld, *De Culturele Factor* (Culemborg, the Netherlands: Lemma, 1988).

14 Overviews of, and introductions to, new institutionalist approaches in political science and sociology are: J. G. March and J. P. Olsen, 'The New Institutionalism: Organizational Factors in Political Life', *American Political Science Review*, vol. 78, 1984, pp. 734–49; *Rediscovering Institutions* (New York: Free Press, 1989); Powell and DiMaggio, op. cit. in note 12; J. Kato, 'Institutions and Rationality in Politics: Three Varieties of Neo-Institutionalists', *British Journal of Political Science*, vol. 26, 1996, pp. 553–82; P. A. Hall and R. C. R. Taylor, 'Political Science and the Three New Institutionalisms', *Political Studies*, vol. 44, 1996, pp. 952–73; E. M. Immergut, 'The Theoretical Core of the New Institutionalism', *Politics and Society*, vol. 26, 1998, pp. 5–34; and B. G. Peters, *Institutional Theory in Political Science* (London: Pinter, 1999).

15 P. J. DiMaggio and W. W. Powell, 'Introduction', in Powell and DiMaggio, op. cit. in note 14.

16 Hall and Taylor, op. cit. in note 14.

17 M. Douglas, *How Institutions Think* (London: Routledge & Kegan Paul, 1987); G. Grendstad and P. Selle, 'Cultural Theory and the New Institutionalism', *Journal of Theoretical Politics*, vol. 7, 1995, pp. 5–27.

18 Following Jepperson, op. cit. in note 12; and F. Hendriks, *Cultural Bias in Policy-Making: An Institutional Comparison of Cities* (Cheltenham: Edward Elgar, 1999).

19 The best introduction to the historical institutionalism is K. Thelen and S. Steinmo, 'Historical Institutionalism in Comparative Politics', in S. Steinmo, K. Thelen and F. Longstreth (eds), *Structuring Politics* (Cambridge: Cambridge University Press, 1992). Some of the main publications are P. J. Katzenstein, *Between Power and Plenty* (Madison, WI: University of Wisconsin Press, 1978); P. A. Hall, *Governing the Economy* (Oxford: Oxford University Press, 1986); Steinmo, Thelen and Longstreth, op. cit.; and S. Steinmo, *Taxation and Democracy: Swedish, British and American Approaches to Financing the Modern State* (New Haven, CT: Yale University Press, 1993).

20 See also H. Eckstein, 'A Cultural Theory of Political Change', *American Political Science Review*, vol. 82, 1988, pp. 789–804.

21 D. J. Puchala and R. F. Hopkins, 'International Regimes: Lessons from Inductive Analysis', in Krasner, op. cit. in note 1.

22 Krasner, op. cit. in note 4, p. 2.

23 D. Snidal, 'The Politics of Scope: Endogenous Actors, Heterogeneity and Institutions', *Journal of Theoretical Politics*, vol. 6, 1994, pp. 451–2.

24 M. Douglas, 'Culture and Collective Action', in M. Freilich (ed.), *The Relevance of Culture* (New York: Bergin & Garvey, 1989); Snidal, op. cit. in note 23; and especially J. Melkin and A. Wildavsky, 'Why the Traditional Distinction between Public and Private Goods Should Be Abandoned', *Journal of Theoretical Politics*, vol. 3, 1991, pp. 355–78.

25 A. O. Hirschman, *Exit, Voice and Loyalty* (Cambridge, MA: Harvard University Press, 1970), p. 101.

26 This part of the text is loosely based on M. H. Banks, 'Four Conceptions of Peace', in D. Sandole and I. Sandole Staroste (eds), *Conflict Management and Problem Solving* (London: Frances Pinter, 1987).

27 These rules are the subject of H. Bull, *The Anarchical Society* (London: Macmillan, 1977), part 2.

28 A theoretical version of this view is J. A. Conybeare, 'International Organization and the Theory of Property Rights', *International Organization*, vol. 34, 1980, pp. 307–34. Daniel Deudney's description of the Philadelphian system of states can serve as a historical example of how individualists would like to structure state relationships. See D. Deudney, 'The Philadelphian State System: Sovereignty, Arms Control, and Balance of Power in the American States–Union circa 1787–1861', *International Organization*, vol. 49, 1995, pp. 191–227. An individualist would also endorse the arguments presented in G. M. Gallarotti, ' The Limits of International Organization: Systematic Failure in the Management of International Relations', *International Organization*, vol. 45, 1991, pp. 183–220.

29 D. Vogel, *Kindred Strangers: The Uneasy Relationship between Politics and Business in America* (Princeton, NJ: Princeton University Press, 1996), p. 32.

30 This suggests that the construction of anarchy as the central problem of world politics in a great many analyses of international relations has a strong normative bent to it. On this see also D. Bigo and J. Y. Haine (eds), *Troubler et Inquiéter: Les Discours du Désordre International* (Paris: L'Harmattan, 1996).

31 The description of the Peace League among Iroquois tribes offered by Neta Crawford is an excellent historical example of how egalitarians would like to structure international relations. N. C. Crawford, 'A Security Regime among Democracies: Cooperation among Iroquois Nations', *International Organization*, vol. 48, 1994, pp. 345–85.

32 The best-known academic expression of this is K. N. Waltz, *Theory of International Politics* (Reading, MA: Addison-Wesley, 1979).

33 This part of the text draws on the alternative views of human nature distinguished in grid-group analysis, see M. Thompson, R. Ellis and A. Wildavsky, *Cultural Theory* (Boulder, CO: Westview Press, 1990), pp. 33–7. These various views of human nature were also discussed in Chapter 2. This part of the text is also based on A. B. Wildavsky and C. Lockhart, 'The Social Construction of Cooperation: Egalitarian, Hierarchical and Individualistic Faces of Altruism', in A. B. Wildavsky, *Culture and Social Theory* (New Brunswick, NJ: Transaction Publishers, 1998).

34 A similar assumption is made in S. K. Chai and A. B. Wildavsky, 'Culture, Rationality, and Violence', in D. J. Coyle and R. J. Ellis (eds), *Politics, Policy, and Culture* (Boulder, CO: Westview Press, 1994). Cf. J. Mercer, 'Anarchy and Identity', *International Organization*, vol. 49, 1995, pp. 229–52.

35 See A. B. Wildavsky, 'Why Self-Interest Means Less outside of a Social Context: Cultural Contributions to a Theory of Rational Choices', *Journal of Theoretical Politics*, vol. 6, 1994, pp. 131–59; and Wildavsky and Lockhart, op. cit. in note 33.

36 Krasner, op. cit. in note 4, p. 2.

37 I base myself on the following applications of cultural theory to the study of environmental issues: M. Thompson and M. Warburton, 'Decision-Making under Contradictory Circumstances: How to Save the Himalayas When You Can't Find out What's Wrong with Them', *Journal of Applied Systems Analysis*, vol. 12, 1985, pp. 3–34; D. J. Coyle, '"This Land Is Your Land, This Land Is My Land": Cultural Conflict in Environmental Regulation and Land-Use Regulation', in Coyle and Ellis, op. cit. in note 34; M. Thompson and A. Trisoglio, 'Managing the Unmanageable', in L. A. Brooks and S. D. Vandeveer (eds), *Saving the Seas: Values, Scientists and International Governance* (College Park, MD: Maryland Sea Grant College, 1997); R. J. Ellis and F. Thompson, 'Culture and the Environment in the Pacific Northwest', *American Political Science Review*, vol. 91, 1997, pp. 885–97; M. Thompson and S. Rayner, 'Risk and Governance, Part I: The Discourses of Climate Change', *Government and Opposition*, vol. 33, 1998, pp. 139–66; and M. Thompson, S. Rayner and S. Ney, 'Risk and Governance, Part II: Policy in a Complex and Plurally Perceived World', *Government and Opposition*, vol. 33, 1998, pp. 331–54.

38 The following book is a forceful expression of the individualist's position on international environmental issues: A. B. Wildavsky, *But Is It True? A Citizen's Guide to Environmental Health and Safety Issues* (Cambridge, MA: Harvard University Press, 1995). In my eyes, Wildavsky's claims about the 'true nature' of environmental problems sit uneasily with the extensive contributions he has made to the development of cultural theory.

39 An effective egalitarian critique of the grand river projects of the past is F. Pearce, *The Dammed: Rivers, Dams, and the Coming of the World Water Crises* (London: Bodley Head, 1992).

40 Such systems are described in L. A. Teclaff and E. Teclaff, 'Restoring River and Lake Basin Eco-Systems', *Natural Resources Journal*, vol. 35, 1995, pp. 905–32.

41 On grid-group theory and risk, see S. Rayner, 'Cultural Theory and Risk Analysis', in S. Krimsky and D. Golding (eds), *Social Theories of Risk* (New York: Praeger, 1992).

42 M. Weber, *Max Weber on Charisma and Institution Building* (Chicago: University of Chicago Press, 1983), pp. 128–9.

43 This is also criticized in H. R. Alker, 'The Presumption of Anarchy in World Politics: On Recovering the Historicity of World Society', in H. R. Alker, *Rediscoveries and Reformulations: Humanistic Methodologies for International Studies* (Cambridge: Cambridge University Press, 1997).

44 P. A. Sabatier, 'Knowledge, Policy-Oriented Learning and Policy Change: An Advocacy Coalition Framework', *Knowledge*, vol. 8, 1987, pp. 649–92.
45 Evans's review of a set of case studies of international negotiations notes that clashing worldviews are a much larger source of conflict than usually recognized in IR theory. P. B. Evans, 'Building an Integrative Approach to International and Domestic Politics: Reflections and Projections', in P. B. Evans, H. K. Jacobson and R. D. Putnam (eds), *Double-Edged Diplomacy* (Berkeley, CA: University of California Press, 1993).
46 M. Schwarz and M. Thompson, *Divided We Stand: Redefining Politics, Technology and Social Choice* (Hemel Hempstead: Harvester-Wheatsheaf, 1990); Hendriks, op. cit. in note 18.
47 A description of RIVM's public participation procedures is M. B. A. van Asselt and J. Rotmans, 'Uncertainty in Perspective', *Global Environmental Change*, vol. 6, 1996, pp. 121–57.

Chapter 4   A Watershed on the Rhine

1 *Le Monde*, 17 October 1996. This article was preceded by a similar one in *The Washington Post*, 27 March 1996, entitled: '"Sewer of Europe" Cleans up its Act.'
2 Ch. Lockhart, 'Political Culture and Political Change', in R. J. Ellis and M. Thompson (eds), *Culture Matters: Essays in Honor of Aaron Wildavsky* (Boulder, CO: Westview Press, 1997).
3 S. D. Krasner, 'Structural Causes and Regime Consequences: Regimes as Intervening Variables', in S. D. Krasner (ed.), *International Regimes* (Ithaca, NY: Cornell University Press, 1983), pp. 3–4.
4 P. A. Hall, 'Policy Paradigms, Social Learning, and the State: The Case of Economic Policymaking in Britain', *Comparative Politics*, vol. 25, 1993, pp. 275–96.
5 G. Osherenko and O. R. Young, 'The Formation of International Regimes: Hypotheses and Cases', in O. R. Young and G. Osherenko (eds), *Polar Politics: Creating International Environmental Regimes* (Ithaca, NY: Cornell University Press, 1993), p. 15.
6 O. A. Young, 'Political Leadership and Regime Formation: On the Development of Institutions in International Society', *International Organization*, vol. 45, 1991, pp. 293–301; O. R. Young and G. Osherenko, 'International Regime Formation: Findings, Research Priorities, and Applications', in Young and Osherenko, op. cit. in note 5, pp. 234–5.
7 E. Adler, 'Cognitive Evolution: A Dynamic Approach for the Study of International Relations and Their Progress', in E. Adler and B. Crawford (eds), *Progress in Postwar International Relations* (New York: Columbia University Press, 1991).
8 His framework is set out in P. A. Sabatier, 'Knowledge, Policy-Oriented Learning and Policy Change: An Advocacy Coalition Framework', *Knowledge*, vol. 8, 1987, pp. 649–92. For empirical applications, as well as theoretical revisions, of this model see P. A. Sabatier and H. C. Jenkins-Smith (eds), *Policy Change and Learning: An Advocacy Coalition Approach* (Boulder, CO: Westview Press, 1993).

9   Sabatier has recently adapted his model slightly under the influence of grid-group theory. See P. A. Sabatier, 'The Advocacy Coalition Framework: Revisions and Relevance for Europe', *Journal of European Public Policy*, vol. 5, 1998, p. 110.

10 See H. Hellmann, 'Load Trends of Selected Chemical Parameters of Water Quality and of Trace Substances in the River Rhine between 1955 and 1988', *Water Science Technology*, vol. 29, 1994, p. 70.

11 This part of the text is based on J. G. Lammers, *Pollution of International Watercourses* (The Hague: Martinus Nijhoff, 1984), pp. 166–95; C. Dieperink, 'Between Salt and Salmon: Network Analysis in the Rhine Catchment Area', in P. Glasbergen (ed.), *Managing Environmental Disputes* (Dordrecht: Kluwer Academic, 1995); and K. Wieriks and A. Schulte-Wülwer-Leidig, 'Integrated Water Management for the Rhine River Basin: From Pollution Prevention to Ecosystem Management', *Natural Resources Forum*, vol. 21, 1997, pp. 147–56.

12 These conventions are both published in *International Legal Materials* (1977), pp. 242–75.

13 In this I somewhat follow K. A. Mingst, 'The Functionalist and Regime Perspectives: The Case of Rhine River Cooperation', *Journal of Common Market Studies*, vol. 20, 1981, pp. 161–73. Her article advocates analysis of the international Rhine regime in terms of the perceptions that are predominant within the involved organizations.

14 Besides the annual reports (entitled *Rheinberichte*) of the IAWR, see Hellmann, op. cit. in note 10; J. E. M. Beurskens, H. J. Winkels, J. de Wolf and C. G. C. Dekker, 'Trends of Priority Pollutants in the Rhine during the Last Fifty Years', *Water Science Technology*, vol. 29, 1994, pp. 77–85; K. G. Malle, 'Der Gütezustand des Rheins', *Chemie in unserer Zeit*, vol. 25, 1991, pp. 257–67; W. M. Stigliani, P. R. Jaffé and S. Anderberg, 'Heavy Metal Pollution in the Rhine Basin', *Environmental Science Technology*, vol. 27, 1993, pp. 786–93.

15 A. Kiss, 'The Protection of the Rhine against Pollution', in A. E. Utton and L. A. Teclaff (eds), *Transboundary Resources Law* (Boulder, CO: Westview Press, 1987), pp. 63–7.

16 'Effluent limit' is a central concept in water protection policies. An effluent limit stands for the maximum amount of a chemical pollutant that the discharge of wastewater by a company or municipality is allowed to contain. This concept should be clearly separated from the notion of 'water quality standard'. A water quality standard denotes either the maximum amount of a particular toxic substance that is allowed to be present in an open water, or the minimum amount of a biological parameter (such as oxygen saturation or biodiversity) that should prevail in an open water. In water protection policies, effluent limits for firms are often calculated on the basis of the water quality standards for the river, lake or sea into which the firms discharge their wastewater.

17 Dieperink, op. cit. in note 11, p. 130.

18 See the EC directives for quicksilver (82/176 and 84/156) and cadmium (83/54). These directives were adopted under the EC Framework Directive for Surface Waters 76/464.

19 A comprehensive analysis of the efforts to reduce salt dumping into

the Rhine is T. Bernauer, 'The International Financing of Environmental Protection: Lessons from Efforts to Protect the River Rhine against Chloride Pollution', *Environmental Politics*, vol. 4, 1995, pp. 369–90.

20 Interviews with officials from the Dutch Ministry of Transport, Public Works and Water Management, The Hague, 14 March 1996; and the *Agence de l'Eau Rhin–Meuse*, Metz, 13 November 1996.

21 Interview with former official of the Dutch Ministry of Foreign Affairs, Amsterdam, 8 April 1996.

22 On these court cases, see H. U. Jessurun d'Oliviera, 'La Pollution du Rhin et le Droit International Privé', in R. Hueting, C. van der Veen, A. C. Kiss and H. U. Jessurun d'Oliviera (eds), *Rhine Pollution* (Zwolle, the Netherlands: Tjeenk Willink, 1978); 'Rijnvervuiling en Internationaal Privaatrecht: Rechtsvergelijkende Aantekeningen', *Milieu en Recht*, vol. 16, 1989, pp. 146–56; and J. M. van Dunné (ed.), *Transboundary Pollution and Liability: The Case of the River Rhine* (Lelystad, the Netherlands: Vermande, 1991).

23 M. T. Kamminga, 'Who Can Clean up the Rhine: The European Community or the International Rhine Commission?', *Netherlands International Law Review*, vol. 25, 1978, pp. 63–9.

24 The German Commission for the Preservation of the Rhine. Representatives from the federal Ministries of the Environment, Economic Affairs, Transport, Agriculture and Foreign Affairs participate in this commission as well as officials from the six *Länder* governments. The Deutsche Kommission zur Reinhaltung des Rheins only takes non-binding decisions on the basis of consensus among its members. The German delegation to the ICPR meetings is drawn from this commission.

25 In the Netherlands: Reinwater; Stichting Natuur en Milieu. In Germany: Greenpeace; World Wide Fund for Nature; Bundesverband Bürgerinitiativen; Naturschutzbund Deutschland. In France: Alsace Nature. In Switzerland: Schweizerischer Bund für Naturschutz. (This list is not exhaustive but only includes names of the major environmental organizations that have been concerned with the protection of the Rhine.)

26 I. Romy, *Les Pollutions Transfrontières des Eaux: L'Exemple du Rhin* (Lausanne: Éditions Paytot, 1990) offers a comprehensive overview of the domestic laws in each Rhine country that have been of relevance to the protection of the Rhine. See also B. Barraqué, 'Les Politiques de l'Eau en Europe', *Revue Française de Science Politique*, vol. 45, 1995, pp. 420–53.

27 In the Netherlands the *Wet Oppervlakte Wateren* of 1970 functioned. In France there was the law 64–125 dating from 1964. In Germany, the *Wasserhaushaltsgesetz* of 1957 was complemented by the *Rheinhaltesordnungen* of the German *Länder*. In Switzerland, the *loi fédérale sur la protection des eaux contre la pollution* of 1972 was in place.

28 Thomas Bernauer and Peter Moser have also concluded that domestic and private measures contributed much more to the clean-up of the Rhine than international agreements. See T. Bernauer and P. Moser, 'Reducing Pollution of the River Rhine: The Influence of International Cooperation', *Journal of Environment and Development*, vol. 5, 1996, pp. 389–415.

29 A. Nollkaemper, 'The River Rhine: From Equal Apportionment to Eco-

system Protection', *Review of European Community and International Environmental Law*, vol. 5, 1996, p. 158; Wieriks and Schulte-Wülwer-Leidig, op. cit. in note 11, p. 155. At the time of writing, Mr Wieriks and Ms Schulte-Wülwer-Leidig were (respectively) Executive Secretary and Deputy Secretary of the ICPR.

30 Interview with official of the Dutch Institute for Inland Water Management and Waste Water Treatment (RIZA), Lelystad, 1 August 1996. See also H. T. A. Bressers and L. A. Plettenburg, 'The Netherlands', in M. Jänicke and H. Weidner (eds), *National Environmental Policies* (Berlin: Springer, 1997), pp. 115–16.

31 M. Jänicke and H. Weidner, 'Germany', in Jänicke and Weidner, op, cit. in note 30, p. 139; W. Rüdig and R. A. Kraemer, 'Networks of Cooperation: Water Policy in Germany', *Environmental Politics*, vol. 3, 1994, pp. 52–79.

32 M. Prieur, *Droit de l'Environnement* (Paris: Dalloz, 1996), pp. 511–13. The French Ministry of the Environment has divided the whole of French territory into six *bassins hydrographiques*. In each of these basins, an *agence de l'eau* is permitted to levy a fee on point source discharges into open waters. The monies collected in this way are invested in water protection programmes by the *agence de l'eau*. These local water agencies are made up of civil servants who fall under the Ministry of the Environment, but who enjoy a relatively large degree of autonomy. The local body relevant for the protection of the Rhine is the *agence de l'eau Rhin–Meuse* in Metz. The French missions to the ICPR include representatives from this water agency as well.

33 S. Schwager, P. Knoepfel and H. Weidner, *Umweltrecht Schweiz-EG: Das schweizerische Umweltrecht im Lichte der Umweltschutzbestimmungen der Europäischen Gemeinschaften – ein Rechtsvergleich* (Basle: Helbing & Lichtenhahn, 1988), p. 20.

34 Adrienne Héritier has shown that something similar has taken place at the EU level. The details of air protection policies in EU member states have differed greatly from one country to the next. Generally, member states have argued for the adoption of their particular regulatory style at the EU level. This insistence by member states on their own way of doing things has sometimes led to impasses in the development of EU environmental regulation. A. Héritier, 'The Accommodation of Diversity in European Policy-Making and Its Outcomes: Regulatory Policy as a Patchwork', *Journal of European Public Policy*, vol. 3, 1996, pp. 149–67.

35 Interview with former official of the ICPR secretariat, Delft, 30 July 1996.

36 The following description of the Sandoz incident is based on A. Schwabach, 'The Sandoz Spill: The Failure of International Law to Protect the Rhine from Pollution', *Ecology Law Quarterly*, vol. 16, 1989, pp. 443–80; and F. Galliot, 'La Coopération Européenne en Matière de Lutte contre la Pollution du Rhin', *Annuaire de Droit Maritime et Aéro-Spatial*, vol. 10, 1989, pp. 247–71.

37 See also A. Nollkaemper, 'The Rhine Action Programme: A Turning Point in the Protection of the North Sea?', in D. Freestone and T. IJlstra (eds),

The North Sea: Perspectives on Regional Environmental Co-operation (London: Graham & Trotman, 1990); Dieperink, op. cit. in note 11, p. 132; C. H. V. de Villeneuve, 'Western Europe's Artery: The Rhine', *Natural Resources Journal*, vol. 36, 1996, pp. 451–3.

38 Respectively, *The Economist*, 15 November 1986; *New York Times*, 21 December 1996, and *Der Spiegel*, 17 November 1986.

39 A 'successful' international regime is of course a highly contestable, and contested, concept. Here an international regime will be marked as successful only when all the stakeholders, adhering to the various ways of life, agree among each other that the regime has indeed been a success. Looking back in 1996, all interviewees (representing environmental groups, industries, agriculture, government) acknowledged that the international regime to protect the Rhine had been hugely effective and successful from 1987 onwards.

40 T. Risse-Kappen, 'Ideas Do Not Float Freely: Transnational Coalitions, Domestic Structures, and the End of the Cold War', in R. N. Lebow and T. Risse-Kappen (eds), *International Relations Theory and the End of the Cold War* (New York: Columbia University Press, 1995).

41 Interview with Ms Kroes, Nijenrode, the Netherlands, 20 August 1996.

42 Interview with former official of Dutch Ministry of Foreign Affairs, Amsterdam, 8 April 1996.

43 Interview with representative of Stichting Reinwater, Amsterdam, 20 March 1996.

44 An example is a publication by Greenpeace members Kerner, Maissen and Radek, which combines a general attack on the profit motive and the chemical industry with virulent criticisms of the integrity of several persons working for Sandoz at the time of the fire. See I. Kerner, T. Maissen and D. Radek, *Der Rhein: Die Vergiftung geht weiter* (Reinbek bei Hamburg: Rowohlt, 1987).

45 See ICPR, *Rhine Action Programme* (Koblenz: 1988).

46 Interview with representative of the German Association of Chemical Firms (VCI), Frankfurt, 10 July 1996.

47 ICPR, *Ecological Master Plan for the Rhine: Salmon 2000* (Koblenz: 1991), p. 9.

48 See ICPR, *The Rhine: An Ecological Revival* (Koblenz: 1995).

49 W. G. Cazemier, 'Present Status of the Salmondis Atlantic Salmon and Sea-Trout in the Dutch Part of the River Rhine', *Water Science Technology*, vol. 29, 1994, pp. 37–41.

50 K. G. Malle, 'Accidental Spills: Frequency, Importance, Control, Countermeasures', *Water Science Technology*, vol. 29, 1994, pp. 149–63.

51 In all riparian countries, water laws and policies were greatly expanded after 1986. Anti-pollution measures were very much tightened, and the principles of ecosystem management were introduced. In the Netherlands, the *Derde Nota Waterhuishouding* was adopted in 1989. In Germany, the following federal laws were passed: the 1987 *Wasch- und Reinigungmittelgesetz*, the 1990 *Abwasserabgabengesetz* and the 1990 *Chemikaliengesetz*. In France, the water law of 3 January 1992 was adopted. In the European Community, the following measures came into force: the Urban

Waste Water Directive 91/271/EEC, and the Nitrates Directive 91/676/
EEC. A comprehensive Ecological Quality of Water Directive, COM (93)
680 final, was proposed by the Commission in 1993, and is still under
consideration by the Council and the Parliament. In Switzerland, the
Rhine Action Programme introduced the theme of ecosystem management in water policy (interview with public servant of the Bundesamt
für Umwelt, Wald und Landwirtschaft, Berne, 14 November 1996).
52 Interview with representative of Directorate-General XI, Unit Water
Protection, European Commission, Brussels, 2 April 1996.
53 See also Wieriks and Schulte-Wülwer-Leidig, op. cit. in note 11, pp.
154–5.
54 Interview with official of the Dutch Ministry of Transport, Public Works,
Water Management, Haarlem, 4 April 1996.
55 Wieriks and Schulte-Wülwer-Leidig, op. cit. in note 11, pp. 152–3; and
Nollkaemper, op. cit. in note 29.
56 An overview of these water quality standards is given in ICPR, *Statusbericht
Rhein* (Koblenz: 1993), pp. 118–20.
57 Respectively R. O. Keohane and J. S. Nye, Jr (eds), *Transnational Relations and World Politics* (Cambridge, MA: Harvard University Press, 1971);
P. M. Haas, R. O. Keohane and M. A. Levy (eds), *Institutions for the
Earth: Sources of Effective International Environmental Protection* (Cambridge,
MA: MIT Press, 1993); A. Chayes and A. Chandler Chayes, 'On Compli-
ance', *International Organization*, vol. 47, 1993, pp. 175–206; R. B. Mitchell,
'Regime Design Matters: Intentional Oil Pollution and Treaty Compli-
ance', *International Organization*, vol. 48, 1994, pp. 425–58; P. M. Haas,
Saving the Mediterranean: The Politics of International Environmental Protection
(New York: Columbia University Press, 1990); and E. B. Haas, *When
Knowledge Is Power* (Berkeley, CA: University of California Press, 1990).
58 Interviews with members of Reinwater, Amsterdam, 20 March 1996;
Amsterdam, 26 March 1996; and Amsterdam, 12 April 1996; World Wide
Fund for Nature – Germany, Rastatt, 11 July 1996; Naturschutzbund
Deutschland, Kranenburg, 12 July 1996; Bundesverband Bürgerinitiativen
Unweltschutz, Freiburg, 11 November 1996; Bund für Naturschutz
Baselland, Basle, 15 November 1996.
59 ICPR, *Grundlagen und Strategie zum Aktionsplan Hochwasser* (Koblenz: 1995),
p. 7.
60 This is based on the accounts and analyses of the floodings of the Rhine
and the Meuse that were published in *Le Monde, Der Spiegel, Frankurter
Allgemeine and NRC Handelsblad* in December 1993 and January 1995.
61 ICPR, op. cit. in note 59.
62 Interview with official of the Dutch Institute for Inland Water Management and Waste Water Treatment (RIZA), Lelystad, 24 January 1997.
63 Personal communication with ICPR official, Koblenz, 8 May 1998.
64 For the case of the Netherlands, see Bressers and Plettenburg, op. cit.
in note 30, p. 116; for Switzerland, see P. Knoepfel, 'Switzerland', in
M. Jänicke and H. Weidner (eds), *National Environmental Policies* (Berlin: Springer, 1997), p. 181; for France, G. Müller-Brandeck-Bocquet,
*Die institutionelle Dimension der Umweltpolitik: Eine vergleichende Untersuchung
zu Frankreich, Deutschland, und der Europäischen Union* (Baden-Baden:

Nomos, 1996), pp. 93–4; for Germany, see Jänicke and Weidner, op. cit. in note 31, pp. 139–40.

65 P. Huisman, 'A Living Rhine Needs More than Water', unpublished manuscript (Delft, the Netherlands: Delft Institute of Technology, no date).

66 For example, F. Pearce, *The Dammed: Rivers, Dams, and the Coming of the World Water Crises* (London: Bodley Head, 1992).

67 D. J. Coyle, '"This Land Is Your Land, This Land Is My Land": Cultural Conflict in Environmental Regulation and Land-Use Regulation', in D. J. Coyle and R. J. Ellis (eds), *Politics, Policy, and Culture* (Boulder, CO: Westview Press, 1994).

68 ICPR, *Topic Rhine* (Koblenz: April 1998), p. 3.

Chapter 5   Who has Washed the River Rhine?

1 Cited in A. Schwabach, 'The Sandoz Spill: The Failure of International Law to Protect the Rhine from Pollution', *Ecology Law Quarterly*, vol. 16, 1989, p. 459.

2 O. A. Young, *International Governance: Protecting the Environment in a Stateless Society* (Ithaca, NY: Cornell University Press, 1994), pp. ix–x.

3 Comparisons between the pollution of the Rhine and other European rivers are K. G. Malle, 'Gewässergüte von Rhein und Elbe – ein Vergleich', *WLB Wasser, Luft und Boden*, vol. 11, 1991, pp. 18–20; 'Systemvergleich Rhein–Wolga', *WLB Wasser, Luft und Boden*, vol. 12, 1992, pp. 20–1.

4 The records of the ICPR are published in its annual reports; the records of the IAWR in its yearly *Rheinberichte*; the data collected by the DKRR in its yearly *Zahlentafeln*. Analyses based on these numbers are H. Hellmann, 'Load Trends of Selected Chemical Parameters of Water Quality and of Trace Substances in the River Rhine between 1955 and 1988', *Water Science Technology*, vol. 29, 1994, pp. 69–76; J. E. M. Beurskens, H. J. Winkels, J. de Wolf and C. G. C. Dekker, 'Trends of Priority Pollutants in the Rhine during the Last Fifty Years', *Water Science Technology*, vol. 29, 1994, pp. 77–85; K. G. Malle, 'Der Gütezustand des Rheins', *Chemie in unserer Zeit*, vol. 25, 1991, pp. 257–67; W. M. Stigliani, P. R. Jaffé and S. Anderberg, 'Heavy Metal Pollution in the Rhine Basin', *Environmental Science Technology*, vol. 27, 1993, pp. 786–93; and T. Tittizer, F. Schöll and M. Dommermuth, 'The Development of the Macrozoobenthos in the River Rhine in Germany during the 20th Century', *Water Science Technology*, vol. 29, 1994, pp. 21–8.

5 A. Nollkaemper, 'The River Rhine: From Equal Apportionment to Ecosystem Protection', *Review of European Community and International Environmental Law*, vol. 5, 1996, pp. 152–60; K. Wieriks and A. Schulte-Wülwer-Leidig, 'Integrated Water Management for the Rhine River Basin: From Pollution Prevention to Ecosystem Management', *Natural Resources Forum*, vol. 21, 1997, pp. 147–56.

6 In 1996, Sandoz and Ciba-Geigy merged to form Novartis AG, at present the world's largest chemical concern.

7 For instance, BASF AG decided to build sewage treatment plants as early

as 1964. K. G. Malle and P. Tonne, 'Die Abwasserreinigung der BASF AG in Ludwigshafen', *Korrespondenz Abwasser*, vol. 39, 1992, p. 534.

8   Interview with representative of German Association of Chemical Firms (VCI), Frankfurt, 10 July 1996.

9   Quoted in Malle and Tonne, op. cit. in note 7, p. 545.

10 The great number of accidental spills into the Rhine in 1986 is catalogued in RIWA, *Jaarverslag 1986, Deel A: De Rijn* (Amsterdam: 1986), pp. 24–34.

11 Interview with official of Dutch Ministry of the Environment, The Hague, 14 March 1996.

12 Figures received from the environment ministries of these *Länder*.

13 Figure received from the German Association of the Chemical Industry (VCI).

14 P. Rogers, *America's Water* (Cambridge, MA: MIT Press, 1996), p. 139.

15 Wieriks and Schulte-Wülwer-Leidig, op. cit. in note 5, p. 153.

16 K. G. Malle, 'De Vermindering van de Vervuiling van de Rijn vanuit het Standpunt der Chemische Industrie', H_2O, vol. 21, 1988, p. 138.

17 Interviews with employees of Shell Pernis BV, Rotterdam, 18 March 1996; Hoechst AG, Frankfurt, 11 July 1996; BASF AG, Ludwigshaven, 24 July 1996; Bayer AG, Leverkussen, 1 August 1996; Sandoz Pharma AG, Basel, 14 November 1996; and Ciba-Geigy AG, 14 November 1996.

18 'WRK' stands for 'Watertransportmaatschappij Rijn-en Kennemerland', the Dutch water supply company that provides the funds for the Rhine award.

19 I. Romy, *Les Pollutions Transfrontières des Eaux: L'Exemple du Rhin* (Lausanne: Éditions Paytot, 1990).

20 Interviews with representatives of the Port of Rotterdam, Rotterdam, 12 April 1996; Municipality of Bonn, Bonn, 11 March 1997; Municipality of Mannheim, Mannheim, 7 April 1997; Municipality of Frankfurt, Frankfurt, 11 April 1997.

21 K. G. Malle, 'Accidental Spills: Frequency, Importance, Control, Countermeasures', *Water Science Technology*, vol. 29, 1994, pp. 149–63.

22 K. G. Malle, 'Cleaning up the River Rhine', *Scientific American*, vol. 274, 1996, p. 57.

23 Ibid., p. 57.

24 Nitrates Directive 91/676/EEC.

25 In 1986, the average amount of nitrates in the Rhine at Lobith equalled 4.36 mg/l. In 1996, this number was 3.59 mg/l. See the year reports of the RIWA for these years (p. 102 and p. 76, respectively).

26 For clashes between these views see D. A. Baldwin (ed.), *Neorealism and Neoliberalism: The Contemporary Debate* (New York: Columbia University Press, 1993); C. W. Kegley, Jr (ed.), *Controversies in International Relations Theory: Realism and the Neoliberal Challenge* (New York: St. Martin's Press, 1995); and J. Grieco, R. Powell and D. Snidal, 'Controversy: The Relative Gains of International Cooperation', *American Political Science Review*, vol. 87, 1993, pp. 729–43.

27 The main neorealist statements are K. N. Waltz, *Theory of International Politics* (Reading, MA: Addison-Wesley, 1979); R. O. Gilpin, *War and Change in World Politics* (Cambridge: Cambridge University Press, 1981);

S. Walt, *The Origins of Alliances* (Ithaca, NY: Cornell University Press, 1987); J. Grieco, *Cooperation among Nations* (Ithaca, NY: Cornell University Press, 1993); J. Mearsheimer, 'The False Promise of International Institutions', *International Security*, vol. 19, 1994/95, pp. 5–49.

28 Some of the main neoliberalist publications are R. Axelrod, *The Evolution of Co-operation* (New York: Basic Books, 1984); R. O. Keohane, *After Hegemony* (Princeton, NJ: Princeton University Press, 1986); O. A. Young, *International Cooperation: Building Regimes for Natural Resources and the Environment* (Ithaca, NY: Cornell University Press, 1989); D. Snidal, 'Relative Gains and the Pattern of International Cooperation', *American Political Science Review*, vol. 85, 1991, pp. 701–26; P. M. Haas (ed.), *Epistemic Communities and International Policy Coordination*, special issue of *International Organization*, vol. 46, 1992; R. O. Keohane and E. Ostrom (eds), *Local Commons and Global Interdependence*, special issue of *Journal of Theoretical Politics*, vol. 6, 1994.

29 R. P. Palan and B. M. Blair, 'On the Idealist Origins of the Realist Theory of International Relations', *Review of International Studies*, vol. 19, 1993, pp. 385–99.

30 O. A. Young, *International Governance: Protecting the Environment in a Stateless Society* (Ithaca, NY: Cornell University Press, 1994).

31 On this problem see J. C. Ribot, 'Market–State Relations and Environmental Policy: Limits of State Capacity in Senegal', in R. D. Lipschutz and K. Conca (eds), *The State and Social Power in Global Environmental Politics* (New York: Columbia University Press, 1993).

32 E. Ostrom, *Governing the Commons* (Cambridge: Cambridge University Press, 1990); E. Ostrom, R. Gardner and J. Walker, *Rules, Games, and Common-Pool Resources* (Ann Arbor, MI: University of Michigan Press, 1994); Keohane and Ostrom, op. cit. in note 28.

33 D. Snidal, 'The Politics of Scope: Endogenous Actors, Heterogeneity and Institutions', *Journal of Theoretical Politics*, vol. 6, 1994, pp. 451–2. See also Chapter 3.

34 W. Blomquist, E. Schlager, S. Y. Tang and E. Ostrom, 'Regularities from the Field and Possible Explanations', in Ostrom, Gardner and Walker, op. cit. in note 32, pp. 301–16.

35 W. Rüdig and R. A. Kraemer, 'Networks of Cooperation: Water Policy in Germany', *Environmental Politics*, vol. 3, 1994, p. 63.

36 Interview with representative of Shell Pernis BV, 18 March 1996.

37 N. Choucri, 'Multinational Corporations and the Global Environment', in N. Choucri (ed.), *Global Accord: Environmental Challenges and International Responses* (Cambridge, MA: MIT Press, 1995); R. Falkner, 'The Roles of Firms in International Environmental Politics*, paper presented at the 37th Annual Convention of International Studies Association (San Diego, CA: April 1996); M. E. Porter and C. van der Linde, 'Toward a New Conception of the Environment–Competitiveness Relationship', *Journal of Economic Perspectives*, vol. 9, 1995, pp. 97–118; and W. Grant, 'Large firms, SMES, Environmental Deregulation and Competitiveness', in U. Collier (ed.), *Deregulation in the European Union: Environmental Perspectives* (London: Routledge, 1997).

38 Choucri, op. cit. in note 37.

39 Falkner, op. cit. in note 37, pp. 29–35.
40 Grant, op. cit. in note 37.
41 Interview with representative of VCI, Frankfurt, 10 July 1996.
42 Falkner, op. cit. in note 37, p. 21.
43 Porter and van der Linde, op. cit. in note 37, pp. 98–104.
44 See T. Bernauer and P. Moser, 'Reducing Pollution of the River Rhine: The Influence of International Cooperation', *Journal of Environment and Development*, vol. 5, 1996, pp. 389–415.
45 Interview with officials from the Dutch Ministry of Agriculture, Nature and Fishery, The Hague, 11 April 1996; and US Department of Agriculture, Madison, Wisconsin, 29 May 1997.
46 J. Esser and W. Fach, 'Crisis Management "Made in Germany": The Steel Industry', in P. J. Katzenstein (ed.), *Industry and Politics in West Germany* (Ithaca, NY: Cornell University Press, 1989), p. 223.
47 This is based on interviews with representatives from the European Chemical Industry Council (CEFIC), Brussels, 3 March 1996; Shell Pernis BV, Rotterdam, 18 March 1996; Verband der Chemischen Industrie (VCI), Frankfurt, 10 July 1996; Hoechst AG, Frankfurt, 11 July 1996; BASF AG, Ludwigshaven, 24 July 1996; Bayer AG, Leverkussen, 1 August 1996; Sandoz AG, Basle, 14 November 1996; Ciba-Geigy AG, Basle, 14 November 1996.
48 By 'large companies' I mean those companies whose size necessitated them to build their own sewage treatment plants.
49 For example, K. G. Malle, 'Verschmutzung des Rheins durch Unfälle', *Spektrum der Wissenschaft* (February 1994), p. 42. Dr Karl-Geert Malle was responsible for water protection at BASF for many years.
50 C. S. Allen, 'Political Consequences of Change: The Chemical Industry', in Katzenstein, op. cit. in note 46, pp. 174–6.
51 Rüdig and Kraemer, op. cit. in note 35, p. 59.
52 This section is based on interviews with representatives of Stichting Reinwater, Amsterdam 20 and 26 March and 12 April 1996; World Wide Fund for Nature – Germany, Rastatt, 11 July 1996; Naturschutzbund Deutschland, Kranenburg, 12 July 1996; Bundesverband Bürgerinitiativen Umweltschutz, Freiburg, 11 November 1996; Bund für Naturschutz Baselland, Basle, 15 November 1996; an interview with the chairperson of Alsace Nature published in *Le Monde*, 17 October 1996; as well as publications by all of these organizations.
53 Rüdig and Kraemer, op. cit. in note 35, pp. 66–8; H. T. A. Bressers, D. Huitema and S. M. M. Kuks, 'Policy Networks in Dutch Water Policy', *Environmental Politics*, vol. 3, 1994, p. 52.
54 This is based on 31 interviews with 34 government officials in all Rhine countries. These officials worked for international commissions, federal ministries, regional governmental organizations (*Länder, agences de l'eau*) and municipalities. See Appendix A.
55 For the case of the Netherlands, see H. T. A. Bressers and L. A. Plettenburg, 'The Netherlands', in M. Jänicke and H. Weidner (eds), *National Environmental Policies* (Berlin: Springer, 1997), p. 116; for Switzerland, see P. Knoepfel, 'Switzerland', in Jänicke and Weidner, op. cit., p. 181; for France, G. Müller-Brandeck-Bocquet, *Die institutionelle Dimension der Umweltpolitik: Eine vergleichende Untersuchung zu Frankreich, Deutschland,*

und der Europäischen Union (Baden-Baden: Nomos, 1996), pp. 93–4; for Germany, see M. Jänicke and H. Weidner, 'Germany', in M. Jänicke and H. Weidner, op. cit., pp. 139–40.

56 Interviews with representatives of the city of Rotterdam, 12 April 1996; the city of Bonn, 11 March 1997; the city of Mannheim, 7 April 1997; and the city of Frankfurt, 11 April 1997.

57 L. A. Patterson, 'Agricultural Policy in the European Community: A Three-Level Game Analysis', *International Organization*, vol. 51, 1997, p. 148; and Bernauer and Moser, op. cit. in note 44, p. 412.

58 D. Baldock and P. Lowe, 'The Development of European Agri-Environmental Policy', in M. Whitby (ed.), *The European Environment and CAP Reform* (Wallingford: CAB International, 1996), p. 11; J. de Vries and T. de Groot, 'Europa en de Nederlandse Landbouw: Een Cultureel-Institutionele Verklaring van Veranderingen in de Landbouwbeleidsgemeenschap', in P. 't Hart, M. Metselaar and B. Verbeek (eds), *Publieke Besluitvorming* (The Hague: VUGA, 1995), p. 321.

59 M. Whitby, 'The Prospect for Agri-Environmental Policies within a Reformed CAP', in Whitby, op. cit. in note 58, p. 227.

60 EC Regulation 2078/92. On this regulation, see Baldock and Lowe, op. cit. in note 58, pp. 18–23.

61 Patterson, op. cit. in note 57, p, 159.

62 de Vries and de Groot, op. cit. in note 58, p. 321.

63 Ibid., p. 322.

64 This section is based on interviews with the Landbouwschap, The Hague, 25 March 1996; Deutscher Bauernverband, Bonn, 10 November 1996; Féderation Nationale des Syndicats d'Exploitants Agricoles, Paris, 17 March 1997. These organizations have represented the agricultural firms in (respectively) the Netherlands, Germany and France. This part of the text is also based on interviews with the Dutch Ministry of Agriculture, Nature and Fishery, The Hauge, 11 April 1996; and the French Ministry of Agriculture, Fishery and Food, Paris, 19 November 1996.

65 Nitrate is life.

66 This view has actually been incorporated in the agri-environmental measures accompanying the 1992 MacSharry reforms.

Chapter 6   Why is the Rhine Cleaner than the Great Lakes?

1 F. Hendriks, *Cultural Bias in Policy-Making: An Institutional Comparison of Cities* (Cheltenham: Edward Elgar, 1999).

2 M. Thompson, 'Waste and Fairness', *Social Research*, vol. 65, 1998, pp. 55–73.

3 A. Lijphart, 'Comparative Politics and the Comparative Method', *American Political Science Review*, vol. 65, 1971, pp. 682–93.

4 S. M. Lipset, *Continental Divide: The Institutions and Values of the United States and Canada* (New Brunswick, NJ: Transaction Publishers, 1990).

5 The Canadian policy initiatives concerning industrial pollution of the Great Lakes have usually explicitly been based on consultation and

cooperation. Examples are the policy initiatives ARET and ARET 2 (Accelerated Reduction and Elimination of Toxics Programs) by Environment Canada, and Great Lakes 2000 by Health Canada.

6 G. King, R. O. Keohane and S. Verba, *Designing Social Inquiry: Scientific Inference in Qualitative Research* (Princeton, NJ: Princeton University Press, 1994), p. 222.

7 The economic links between the US Great Lakes states and Ontario, the Canadian province in the basin, are spelled out in D. R. Allardice and W. A. Testa, 'Binational Economic Linkages within the Great Lakes Region', in W. A. Testa (ed.), *The Great Lakes Economy* (Chicago: Federal Reserve Bank of Chicago, 1991).

8 G. Hoberg, 'Sleeping with an Elefant: The American Influence on Canadian Environmental Protection', *Journal of Public Policy*, vol. 10, 1991, pp. 107–32.

9 The information in this section is largely based on the Government of Canada and US Environmental Protection Agency, *The Great Lakes: An Environmental Atlas and Resource Book* (Toronto and Chicago: 1995).

10 E. H. Erdevig, 'An Overview of the Economy of the Great Lakes States', in Testa, op. cit. in note 7, p. 26.

11 Information received from the Great Lakes Fishery Commission in Ann Arbor, Michigan.

12 EPA, 1997. These publications are based on the wastewater data reported by Great Lakes firms. Such public reporting by firms is obligatory in the United States under the 1986 Emergency Planning and Community-Right-to-Know Act.

13 This is explained in K. Wieriks and A. Schulte-Wülwer-Leidig, 'Integrated Water Management for the Rhine River Basin: From Pollution Prevention to Ecosystem Management', *Natural Resources Forum*, vol. 21, 1997, p. 152. At the time of writing Mr. Wieriks and Ms Schulte-Wülwer-Leidig were, respectively, the Executive Secretary and the Deputy Secretary of the ICPR.

14 Samenwerkende Rijn- en Maaswaterleidingbedrijven (RIWA), *Jaarverslag 1996, Deel A: De Rijn* (Amsterdam, 1996), p. 14.

15 C. S. Allen, 'Political Consequences of Change: The Chemical Industry', in P. J. Katzenstein (ed.), *Industry and Politics in West Germany* (Ithaca, NY: Cornell University Press, 1989), p. 172.

16 J. P. Manno, 'Advocacy and Diplomacy: NGOs and the Great Lakes Water Quality Agreement', in T. Princen and M. Finger (eds), *Environmental NGOs in World Politics* (London: Routledge, 1995), p. 72.

17 P. M. Haas, *Saving the Mediterranean: The Politics of International Environmental Protection* (New York: Columbia University Press, 1990).

18 G. R. Francis and H. A. Regier, 'Barriers and Bridges to the Restoration of the Great Lakes Basin Ecosystem', in C. S. Holling and S. S. Light (eds), *Barriers and Bridges to the Renewal of Ecosystems and Institutions* (New York: Columbia University Press, 1995), p. 271.

19 Documented in Francis and Regier, op. cit. in note 18.

20 I. Romy, *Les Pollutions Transfrontières des Eaux: L'Exemple du Rhin* (Lausanne: Éditions Paytot Lausanne, 1990); G. Müller-Brandeck-Bocquet, *Die institutionelle Dimension der Umweltpolitik: Eine vergleichende Untersuchung*

zu Frankreich, Deutschland, und der Europäischen Union (Baden-Baden: Nomos, 1996); H. T. A. Bressers and L. A. Plettenburg, 'The Netherlands', in M. Jänicke and H. Weidner (eds), *National Environmental Policies* (Berlin: Springer, 1997); M. Jänicke and H. Weidner, 'Germany', in Jänicke and Weidner, op. cit.; P. Knoepfel, 'Switzerland', in Jänicke and Weidner, op. cit.; R. N. L. Andrews, 'United States', in M. Jänicke and Weidner, op. cit.

21 An effluent limit stands for the maximum amount of a chemical pollutant that wastewater is allowed to contain. A standard denotes either the maximum amount of a particular toxic substance that is allowed in an open water, or the minimum amount of a biological parameter (such as biodiversity) that should prevail in an open water.

22 D. Vogel, *National Styles of Regulation: Environmental Policy in Great Britain and the United States* (Ithaca, NY: Cornell University Press, 1986), p. 162.

23 In establishing the toxicity of the effluents of the Rhine companies I have relied on measurements of the Rhine's water quality (see Table 6.1). However, as I argued above, the measurement systems for the water quality of the Rhine are so fine-grained as to make these measurements a very good indicator of industrial effluents. This proxy is not in any way influenced by the river's natural conditions. For the Great Lakes case, I have relied on the obligatory public reports by firms of their effluents, as summarized by the EPA for the Great Lakes area.

24 For example, D. Vogel, op. cit. in note 22, p. 159.

25 For example, O. Renn and R. Finson, *The Great Lakes Clean-up Program: A Role Model for International Cooperation?*, European University Institute Working Paper (EPU), No. 91/7 (San Domenico di Fiesole, Italy: European University Institute, 1991).

26 Vogel, op. cit. in note 22, pp. 164–6; R. W. Adler, J. C. Landman and D. M. Cameron, *The Clean Water Act: 20 Years Later* (Washington, DC: Island Press, 1993).

27 V. O. Okaru, 'Financing Publicly Owned Treatment Works and Instituting Enforcement Measures against Non-Compliant Works under the Clean Water Act', *Buffalo Environmental Law Review*, vol. 2, 1994, p. 215.

28 This part of the text is mainly based on three sources of data: (a) the interviews held with stakeholders – see Appendix B; (b) D. R. Allardice, R. H. Mattoon and W. A. Testa, 'Industry Approaches to Environmental Policy in the Great Lakes Region', *University of Toledo Law Review*, vol. 25, 1994, pp. 357–77; and (c) two hearings of the US Senate on proposed amendments of water acts. The first of these is *Amending the Clean Water Act: Hearings before the Subcommittee on Environmental Pollution*, 27 and 28 March 1985. The second is *Water Pollution Prevention and Control: Hearing before the Subcommittee on Environmental Protection*, 21 May; 13 June; 9, 17 and 18 July 1991. During both hearings a number of Great Lakes stakeholders (firms as well as environmental organizations) testified.

29 This is again based on the three sources of data mentioned in note 28.

30 This book is T. A. Colborn, D. Dumanoski and J. Peterson Myers, *Our Stolen Future: Are We Threatening our Fertility, Intelligence and Survival?*

(New York: Plume, 1997). For her work on Great Lakes matters see T. A. Colborn, A. Davidson, S. N. Green, R. A. Hodge, C. I. Jackson and R. A. Liroff, *Great Lakes, Great Legacy?* (Baltimore, MD: Conservation Foundation, 1990).

31 S. Lustig and U. Brunner, 'Environmental Organizations in a Changing Environment – Major US Environmental Organizations in the 1990s', *Environmental Politics*, vol. 5, 1996, p. 131.

32 D. J. Coyle, '"This Land Is Your Land, This Land Is My Land": Cultural Conflict in Environmental Regulation and Land-Use Regulation', in D. J. Coyle and R. J. Ellis (eds), *Politics, Policy, and Culture* (Boulder, CO: Westview Press, 1994).

33 *International Herald Tribune*, 9 March 1998, p. 8: 'Is Cutting US Immigration the Way to Save the Planet?'

34 An overview of important court cases involving the implementation of the Clean Water Act can be found in P. A. Evans (ed.), *The Clean Water Act Handbook* (Chicago: American Bar Association, 1993), pp. 259–60.

35 Following R. Jepperson, 'Institutions, Institutional Effects, and Institutionalism', in W. W. Powell and P. J. DiMaggio (eds), *The New Institutionalism in Organizational Analysis* (Chicago: University of Chicago Press, 1991); and Hendriks, op. cit. in note 1.

36 For an introduction to this literature see K. D. McRae, 'Contrasting Styles of Democratic Decision-Making: Adversarial versus Consensual Politics', *International Political Science Review*, vol. 18, 1997, pp. 279–95.

37 G. J. Ikenberry, 'Conclusion: An Institutional Approach to American Foreign Economic Policy', *International Organization*, vol. 42, 1988, p. 226.

38 S. M. Lipset, *American Exceptionalism: A Double-Edged Sword* (New York: W. W. Norton, 1996), p. 31.

39 Besides A. de Tocqueville, *Democracy in America: Part 1* (New York: Alfred A. Knopf, 1991/1835), some main publications on American exceptionalism are L. Hartz, *The Liberal Tradition in America: An Interpretation of American Political Thought since the Revolution* (New York: Brace & World, 1955); S. P. Huntington, *American Politics: The Promise of Disharmony* (Cambridge, MA: Belknap Press, 1981); B. E. Shafer (ed.), *Is America Different? A New Look at American Exceptionalism* (Oxford: Clarendon Press, 1991); R. J. Ellis, *American Political Cultures* (Oxford: Oxford University Press, 1993); D. J. Elazar, *The American Mosaic: The Impact of Space, Time and Culture on American Politics* (Boulder, CO: Westview Press, 1994); and Lipset, op. cit. in note 38.

40 Vogel, op. cit. in note 22, pp. 253–4; A. B. Wildavsky, 'Resolved, That Individualism and Egalitarianism Be Made Compatible in America: Political-Cultural Roots of Exceptionalism', in B. E. Shafer (ed.), *Is America Different? A New Look at American Exceptionalism* (Oxford: Clarendon Press, 1991); R. A. Kagan, 'Do Lawyers Cause Adversarial Legalism? A Preliminary Inquiry', *Law and Social Inquiry*, vol. 19, 1994, pp. 1–62.

41 Huntington, op. cit. in note 39, p. 56.

42 Vogel, op. cit. in note 22, pp. 29 and 48, respectively.

43 See the Council of the Great Lakes Industries, *Annual Report* (Ann Arbor, MI: 1996); *Report of the Second Roundtable of North American Energy Policy* (Ann Arbor, MI, 1997). Also interview materials.

44 Spelled out in C. B. Blankart, 'A Public-Choice View of Swiss Liberty', *Publius*, vol. 23, 1993, pp. 83–96.

45 P. J. Katzenstein, *Corporatism and Change: Austria, Switzerland, and the Politics of Industry* (Ithaca, NY: Cornell University Press, 1984), p. 110.

46 G. K. Wilson, 'Why Is There No Corporatism in the US?', in G. Lehmbruch and P. C. Schmitter (eds), *Patterns of Corporatist Policy-Making* (London: Sage, 1982), p. 162.

47 E. Blankenburg, 'Changes in Political Regimes and the Continuity of the Rule of Law in Germany', in H. Jacob, E. Blankenburg, H. M. Kritzer, D. M. Province and J. Sanders, *Courts, Law, and Politics in Comparative Perspective* (New Haven, CT: Yale University Press, 1996); E. Blankenburg and F. Bruinsma, *Dutch Legal Culture* (Deventer: Kluwer, 1991); K. M. Holland, 'The Courts in the Federal Republic of Germany', in J. C. Waltman and K. M. Holland (eds), *The Political Role of Law Courts in Modern Democracies* (New York: St. Martin's Press, 1988); 'The Courts in the United States', in Waltman and Holland, op. cit.; H. Jacob 'Courts and Politics in the United States', in Jacob et al., op. cit.; Knoepfel, op. cit. in note 20; D. M. Province, 'Courts in the Political Process in France', in Jacob et al., op. cit.; and D. Radamaker, 'The Courts in France', in Waltman and Holland, op. cit.

48 Blankenburg, op. cit. in note 47.

49 M. Galanter, 'The Assault on Civil Justice: The Anti-Laywer Dimension', in L. M. Friedman and H. N. Scheiber (eds), *Legal Culture and the Legal Profession* (Boulder, CO: Westview Press, 1996), pp. 97–100.

50 For example, *Reynold Metals Company and United States Brewers Association* v. *Environmental Protection Agency*, 760 F.2d 549 (US App. DC, 1985); *American Iron and Steel Institute* v. *Environmental Protection Agency*, 325 (US App. DC 76, 1997); *American Forest and Paper Association* v. *Environmental Protection Agency*, 137 (US App. DC, 1998).

51 For example, *Ford Motor Company* v. *Environmental Protection Agency*, 718 F.2d 55 (US App., 1983); *Mobil Oil Corporation* v. *Environmental Protection Agency*, 716 F.2d 1187 (US App., 1983); *Cerro Copper Products Company* v. *William D. Ruckelshaus, Adminstrator, EPA*, 766 F.2d 1060 (US App., 1985); *South Holland Metal Finishing Company* v. *Carol Browner, Administrator, EPA*, 97 F.3d (US App., 1996); *Great Lakes Chemical Corporation* v. *Environmental Protection Agency* (US App., 1999); *Spitzer Great Lakes Ltd.* v. *Environmental Protection Agency*, 173 F.3d 412 (US App., 1999).

52 For example, *Citizens for a Better Environment et al.* v. *EPA et al.*, 231 (US App. DC 79, 1983); *Friends of the Chrystal River et al.* v. *EPA* 794 F. Supp. 674 (US Dist. LEXIS 11947; 1992); *Sierra Club et al.* v. *EPA*, 843 F. Supp. 1304 (US Dist. LEXIS 19126, 1993); *The Raymond Proffitt Foundation* v. *EPA*, 930 F. Supp. 1088 (US Dist. LEXIS 4872, 1996); *National Wildlife Federation et al.* v. *EPA*, 936 F. Supp. 435 (US Dist. LEXIS 8992, 1996).

53 For example, *Atlantic States Legal Foundation* vs. *Universal Tool & Stamping Co.*, 786 F. Supp. 743 (US Dist. LEXIS 2942, 1992); *Atlantic States Legal Foundation* v. *Eastman Kodak Co.*, 12 F.3d 353 (US App. LEXIS 32711, 1993); *Atlantic States Legal Foundation* v. *Stroh Die Casting Co.*, 116 F.3d 814 (US Dist. LEXIS 14526, 1997).

54 A. Lijphart, 'Introduction', in A. Lijphart (ed.), *Parliamentary versus Presidential Government* (Oxford: Oxford University Press, 1992).
55 R. Brickman, S. Jasanoff and T. Ilgen, *Controlling Chemicals: The Politics of Regulation in Europe and the United States* (Ithaca, NY: Cornell University Press, 1985), p. 72.
56 Allardice, Mattoon and Testa, op. cit. in note 28.
57 G. K. Wilson, *The Politics of Safety and Health: Occupational Safety and Health in the United States and Britain* (Oxford: Clarendon Press, 1985), p. 162.
58 See for instance the testimonies of various Great Lakes firms and environmental organizations during two hearings of the US Senate on proposed amendments of water laws: *Amending the Clean Water Act: Hearings before the Subcommittee on Environmental Pollution*, 27–28 March 1985; and: *Water Pollution Prevention and Control: Hearing before the Subcommittee on Environmental Protection*, 21 May – 18 July, 1991.
59 Lijphart, op. cit. in note 54, pp. 5–6.
60 Ibid., p. 8.
61 Müller-Brandeck-Bocquet, op. cit. in note 20, pp. 39–40.
62 P. C. Schmitter, 'Still the Century of Corporatism?', in P. C. Schmitter and G. Lehmbruch (eds), *Trends toward Corporatist Intermediation* (London: Sage, 1979), p. 13.
63 Schmitter, op. cit. in note 62, p. 15.
64 P. C. Schmitter, 'Corporatism Is Dead! Long Live Corporatism!', *Government and Opposition*, vol. 24, 1989, pp. 54–73; P. C. Schmitter and J. R. Grote, 'Der korporatistische Sisyphus: Vergangenheit, Gegenwart, und Zukunft', *Politische Vierteljahresschrift*, vol. 38, 1997, pp. 530–54; G. Lehmbruch, 'Der Beitrag der Korporatismusforschung zur Entwicklung der Steuerungstheorie', *Politische Vierteljahresschrift*, vol. 37, 1996, pp. 735–51.
65 G. Lehmbruch, 'Consociational Democracy and Corporatism in Switzerland', *Publius*, vol. 23, 1993, pp. 43–60; Latzenstein, op. cit. in note 45; and P. J. Katzenstein (ed.), *Industry and Politics in West Germany* (Ithaca, NY: Cornell University Press, 1989); A. Lijphart and M. M. L. Crepaz, 'Corporatism and Consensus Democracy in Eighteen Countries: Conceptual and Empirical Linkages', *British Journal of Political Science*, vol. 21, 1991, pp. 235–46. Swiss corporatism has been a special case as it has functioned without a strong central government; see H. Kriesi, 'Federalism and Pillarization: The Netherlands and Switzerland Compared', *Acta Politica*, vol. 31, 1996, p. 540.
66 Lijphart and Crepaz, op. cit. in note 65, p. 238.
67 Allen, op. cit. in note 15, pp. 174–6.
68 Wilson, op. cit. in note 46.
69 Schmitter, op. cit. in note 62, p. 15.
70 R. A. Dahl, *A Preface to Democratic Theory* (Chicago: University of Chicago Press, 1956).
71 The above-mentioned effects of corporatism on the protection of the Rhine have therefore only been *indirect*: through familiarizing business leaders with the idea that firms also have social responsibilities, and through creating professional associations strong enough to influence and coordinate the environmental policies of firms.

72　Knoepfel, op. cit. in note 20, p. 180; Jänicke and Weidner, op. cit. in note 20, p. 139; Bressers and Plettenburg, op. cit. in note 20, pp. 115–16; G. Müller-Brandeck-Bocquet, op. cit. in note 20, pp. 37–8.

73　Andrews, op. cit. in note 20, pp. 28–9. On the EPA's handling of Great Lakes firms, see Allardice, Mattoon and Testa, op. cit. in note 28, pp. 357–8.

74　Brickman, Jasanoff and Ilgen, op. cit. in note 55, p. 75.

75　Interview materials.

76　Vogel, op. cit. in note 22, p. 163: 'Violators of the government's water-pollution standards . . . could be fined up to $25 000 a day and sentenced to one year in prison.'

77　See, for instance, *United States* v. *Bethlehem Steel Corporation* (US District Lexis 10482, 1993); *United States* v. *Great Lakes Castings Corporation* (US District Lexis 5745, 1994); *United States* v. *GK Technologies and Indiana Steel & Wire Co.* (US District Lexis 3783, 1997).

78　Bressers and Plettenburg, op. cit. in note 20, p. 116; Knoepfel, op. cit. in note 20, p. 181.

79　Bressers and Plettenburg, op. cit. in note 20, p. 116.

80　Müller-Brandeck-Bocquet, op. cit. in note 20, pp. 62 and 73–87.

81　H. Jessurun d'Oliviera, 'Rijnvervuiling en Internationaal Pivaatrecht: Rechtsvergelijkende Aantekeningen', *Milieu en Recht*, vol. 16, 1989, pp. 146–56.

82　H. Kriesi, R. Koopmans, J. W. Duyvendak and M. Giugni, 'New Social Movements and Political Opportunities', *European Journal of Political Research*, vol. 22, 1992, pp. 219–44.

83　D. Vogel, *Kindred Strangers: The Uneasy Relationship between Politics and Business in America* (Princeton, NJ: Princeton University Press, 1996), p. 319.

84　L. C. Thurow, *Head to Head* (New York: William Morrow, 1992); J. A. Hart, *Rival Capitalists* (Ithaca, NY: Cornell University Press, 1992); M. Albert, *Capitalism versus Capitalism* (New York: Four Walls Eight Windows, 1993); R. J. Hollingsworth and R. Boyer, *Contemporary Capitalism* (Cambridge: Cambridge University Press, 1997); R. Boyer, 'The Diversity and Future of Capitalisms', in G. Hodgson and N. Yokokawa (eds), *Capitalism in Evolution* (Cheltenham: Edward Elgar, forthcoming).

85　R. Liedtke, *Wem gehört die Republik?* (Frankfurt am Main: Eichborn, 1999).

86　An exception to these practices is usually made for the working conditions within US multinationals.

87　International Joint Commission, *Sixth Biennial Report on Great Lakes Water Quality* (Windsor, Ontario: 1992), p. 57.

88　Overviews are T. M. Moe, 'The New Economics of Organization', *American Journal of Political Science*, vol. 28, 1984, pp. 739–77; and B. Weingast, 'Political Institutions: Rational Choice Perspectives', in R. Goodin and H. D. Klingenmann (eds), *A New Handbook of Political Science* (Oxford: Oxford University Press, 1996).

89　M. D. McCubbins, R. G. Noll and B. R. Weingast, 'Administrative Procedures as Instruments of Political Control', *Journal of Law, Economics and Organization*, vol. 3, 1987, pp. 243–77.

90　M. D. McCubbins and T. Schwartz, 'Congressional Oversight Overlooked: Police Controls versus Fire Alarms', in M. D. McCubbins and T. Sulivan (eds), *Congress: Structure and Policy* (Cambridge: Cambridge University Press, 1987).

91   For example, K. A. Shepsle and B. R. Weingast, 'The Institutional Foundations of Committee Power', *American Political Science Review*, vol. 81, 1987, pp. 85–104; M. D. McCubbins, R. G. Noll and B. R. Weingast, 'Structure and Process, Policy and Politics: Administrative Arrangements and the Political Control of Agencies', *Virginia Law Review*, vol. 75, 1989, pp. 431–82; J. A. Ferejohn and C. Shipan, 'Congressional Influence on Bureaucracy', *Journal of Law, Economics and Organization*, vol. 6, 1990, pp. 1–20.

92   Vogel, op. cit. in note 22; J. Badaracco, *Loading the Dice: A Five-Country Study of Vinyl Chloride Regulation* (Cambridge, MA: Harvard Business School Press, 1985); Wilson, op. cit. in note 57; and Brickman, Jasanoff and Ilgen, op. cit. in note 55.

93   Andrews, op. cit. in note 20, p. 27.

94   See the Council of the Great Lakes Industries, op. cit. in note 43; and Allardice, Mattoon and Testa, op. cit. in note 28, pp. 371–3.

Chapter 7   Conclusion: Cultures and Institutions Both Matter in Transboundary Relations

1   Here I am making use of the assumption that ways of life can be distinguished at various levels of analysis – see Chapter 2.

2   A. B. Wildavsky, 'Resolved, That Individualism and Egalitarianism Be Made Compatible in America: Political-Cultural Roots of Exceptionalism', in B. E. Shafer (ed.), *Is America Different? A New Look at American Exceptionalism* (Oxford: Clarendon Press, 1991); R. J. Ellis, *American Political Cultures* (Oxford: Oxford University Press, 1993); D. J. Elazar, *The American Mosaic: The Impact of Space, Time and Culture on American Politics* (Boulder, CO: Westview Press, 1994).

3   R. D. Putnam, *Making Democracy Work: Civic Traditions in Modern Italy* (Princeton, NJ: Princeton University Press, 1993).

Index

Abwassertechnische Vereinigung, 125–6
Adler, Emanuel, 76
Administrative Procedure Act (US, 1946), 168
administrative procedures, 203–4
administrative review, 188–9
adversarial relationships and processes, 46, 158, 176, 184–9, 196–208, 213, 215
agent-structure problem, 30, 37
Agricultural Environmental Regulation, 147
agricultural sector, 110, 112, 121, 124, 131–3
  solidarity with and within, 147–8
  way of life of, 144–9, 217
air pollution, 119, 142, 160
Alexander, Jeffrey, 5
Allen, Christopher, 167
alluvial plains, 100–2, 105, 112
Almond, Gabriel, 5, 36
American Automobile Manufacturers Association, 175, 190
American Forest and Paper Association, 165, 190
anarchy, 56–7, 63–6, 122
  cultures of, 4
  definition of, 82
Andrews, Richard, 207
'Anglo-Saxon capitalism', 199
animal species, disappearance of, 98–9, 160–1
Atlantic States Legal Foundation, 190
autarchy, 55
authority rules, 125–6

Badaracco, J., 206
banks, role of, 199
Barnett, Michael, 4
Basle, 90–1

Bayer AG, 118, 200
belief systems, 38, 63
Bern Convention on the Protection of the Rhine (1976), 80
Bernauer, Thomas, 130–2
'best available technology' approach to environmental improvement, 138–9, 171, 177–8
Biersteker, Thomas, 1–2
bioaccumulation, 173–4
'Bio-Hochreaktor', 118, 200
black-listed substances, 83, 86, 170
boundary rules, 125
Boundary Waters Treaty (1909), 161
Brickman, R., 206
brokerage of policy, 77, 92, 106
Bundesverband Bürgerinitiativen Umweltschutz, 94
business associations, 165, 175, 189–94
business–government relations, 193, 198

Canada, 158, 165, 170
Canada-United States Inter-University Seminars for the Great Lakes, 169
Canadian Chlorine Coordinating Committee, 202
canalization, 79, 100, 112, 141
capitalism, 140, 179
  rival forms of, 199–201
Central Commission for Navigation on the Rhine (CCNR), 79–80
central planning, 56, 64–5
Checkel, Jeffrey, 3–4, 35
chemical discharges, 60, 62, 137, 140–41
  see also, toxic discharges
chemical industry, 113, 118, 124–5, 128, 131, 133, 167